Solutions Manual
for
Introduction to Genetic Analysis
Ninth Edition

WILLIAM D. FIXSEN AND DIANE K. LAVETT

Exploring Genomes:
Web-Based Bioinformatics Tutorials

PAUL G. YOUNG

W. H. Freeman and Company
New York

ISBN-13: 978-1-4292-0177-3
ISBN-10: 1-4292-0177-0

Printed in the United States of America

Fourth printing

W. H. Freeman and Company
41 Madison Avenue
New York, NY 10010
Houndmills, Basingstoke RG21 6XS, England

www.whfreeman.com

Contents

1

The Genetic Approach to Biology

BASIC PROBLEMS

1. *Genetics* is the study of genes and genomes: their biochemical basis; how they function; how they are controlled; how they are organized; how they replicate; how they change; how they can be manipulated; and how they are transmitted from cell to cell and generation to generation.

   The ancient Egyptian racehorse breeders can be only loosely classified as geneticists because their interests were highly focused on producing fast horses, rather than on attempting to understand the mechanisms of heredity. Their understanding of the processes involved in producing fast horses was very incorrect and their methods were not analytic in the modern sense of the word. Nevertheless, they did produce very fast horses through a combination of observation, trial and error, and artificial selection.

2. DNA determines all the specific attributes of a species (shape, size, form, behavioral characteristics, biochemical processes, etc.) and sets the limits for possible variation that is environmentally induced.

3. Properties of DNA that are vital to its being the hereditary molecule are: its ability to replicate; its informational content; and its relative stability while still retaining the ability to change or mutate. Alien life forms might utilize RNA, just as some viruses do, as a hereditary molecule.

4. There are four possible nucleotide pairs at each position (A–T, T–A, G–C, or C–G). Therefore, the general formula for calculating the different possibilities is 4^n (where n = the number of pairs). In this case, there are 4^{10} or 1,048,576 possible DNA molecules!

5. If the DNA is double stranded, A = T and G = C and A + T + C + G = 100%. If T = 15% then C = [100 − 15(2)]/2 = 35%.

6. If the DNA is double stranded, G = C = 24% and A = T = 26%.

7. Human somatic cells are diploid ($2n$) and the haploid number is 23 ($n = 23$). Each chromosome is a double-stranded DNA molecule, so there are 46 molecules of DNA. It is generally stated that the haploid number represents the number of different types of DNA molecules but that does not take into account the difference between the X and Y chromosome in males. So females have 23 different types (called 1–22 and X) and males have 24 (1–22, X, and Y).

8. Keeping this strand antiparallel to the strand given, the sequence would be

T A A C C A C G T A A T G A A G T C C G A G A

9. Yes. There are no sequence restrictions in single-stranded DNA. The percentage of A must equal the percentage of T only in double-stranded DNA.

10. **a.** Yes. Because A = T and G = C, the equation A + C = G + T can be rewritten as T + C = C + T by substituting the equal terms.

b. Yes. The percentage of purines will equal the percentage of pyrimidines in double-stranded DNA.

11. For the following, normal typeface represents previously polymerized nucleotides and *italic* typeface represents newly polymerized nucleotides.

5´ − T T G G C A C G T C G T A A T − 3´
3´ − *A A C C G T G C A G C A T T A* − 5´

5´ − *T T G G C A C G T C G T A A T* − 3´
3´ − A A C C G T G C A G C A T T A − 5´

12. Remember, the transcript will be antiparallel to the DNA template.

5´ — U U G G C A C G U C G U A A U — 3´

13. There are several outcomes possible depending on the exact cause of the null allele. For the representation below, it is assumed that the null allele is still transcribed and translated but that the protein product of the allele is altered in a way that affects its mobility within the gel. (For example, it is slightly smaller and migrates to a position that is lower.) For both gels, lane 1 represents two normal alleles, lane 2 represents one normal and one null allele, and lane 3 represents two null alleles. One other possibility is that the mutant allele is neither transcribed nor translated, in which case lane 3 of both blots would be "blank."

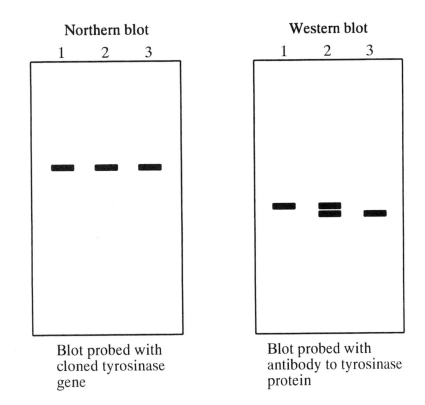

Northern blot

1   2   3

Blot probed with
cloned tyrosinase
gene

Western blot

1   2   3

Blot probed with
antibody to tyrosinase
protein

14. The simplest definition is that a *gene* is a chromosomal region capable of making a functional transcript. However, this does not take into account the regulatory regions near the gene necessary for the proper expression of the gene nor the regions that help control transcription that can be quite distant. Also, many eukaryotes have large regions of noncoding sequences (introns) interspersed within the regions that encode the product (exons.)

15. Electrophoresis separates DNA molecules by size. When DNA is carefully isolated from *Neurospora* (which has seven different chromosomes) seven bands should be produced using this technique. Similarly, the pea has seven different chromosomes and will produce seven bands (homologous chromosomes will co-migrate as a single band). The housefly has six different chromosomes and should produce six bands.

16. The anticodon associates with the codon in an antiparallel orientation. Thus, to pair with the codon 5´—UUA—3´, the anticodon must be 3´—AAU—5´.

17. **a.** For the fungus to be orange (as in mutant 1), orange pigment would have to accumulate and then not be converted into red pigment. This would occur if enzyme C was defective.

b. Mutant 2 is yellow. Using similar logic as in part (a), enzyme B must be defective.

c. An organism defective in both enzyme B and C would still have the phenotype associated with the block at the earlier step of this biochemical pathway. In this case, it would be yellow due to the lack of enzyme B.

CHALLENGING PROBLEMS

18.   *Unpacking the problem*

1.   *Lathyrus odoratus* (sweet peas) are grown for their fragrant, colorful flowers and in fact, their seeds are poisonous. The edible pea belongs to a different but related genus (*Pisum*).

2.   A pathway is a set of biochemical reactions that begins with a precursor molecule and through a series of steps, converts it into some product. In this case, a precursor molecule is converted into an anthocyanin pigment.

3.   There are two pathways described.

4.   Yes. As described, the two pathways are independent.

5.   In biology, a pigment is any molecule that reflects or transmits specific wavelengths of light — its color depends on which wavelengths of light are absorbed or reflected.

6.   In this case, a plant unable to synthesize pigment would be "colorless" or white. Many common solutes are colorless, such as salt (NaCl) and sucrose (table sugar).

7.   As stated above, white.

8.   Yes. The almost infinite variety of paint colors are obtained by mixing a very few primary pigments into a white base. In sweet peas, purple is achieved through the mixing of red and blue pigments.

9.   For now, a mutation is any alteration of the DNA that leads to a different and distinct variant, compared to the more common "normal" phenotype.

10. A null mutation is a change in the DNA that completely prevents the production of a functional product (in this case, enzyme).

11. Null mutations may arise through many types of changes: small deletions or duplications leading to frameshift mutations; nucleotide-pair substitutions that destroy the active site, lead to premature stop codons, or affect proper splicing (in eukaryotes); insertions of DNA; etc.

12. The organism is diploid. It has two copies of each of its chromosomes and therefore two copies of each of its genes.

13. The enzymes encoded by the genes are proteins. Whether or not the proteins are properly encoded (and the enzymes active) is the basis of this analysis.

14. No. The genetic locations of these genes are not relevant to this problem.

   In the following schematics, the "nucleus" separates the DNA and its transcription from translation. Also, the "defect" within the gene that results in the production of non-functional enzyme is marked by an "x."

15.

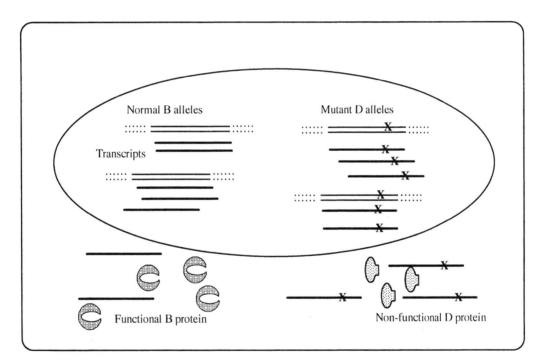

16.

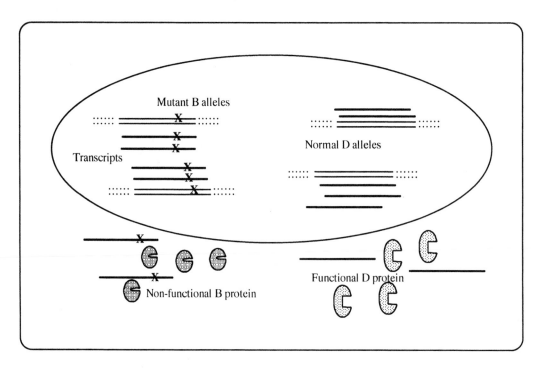

17.

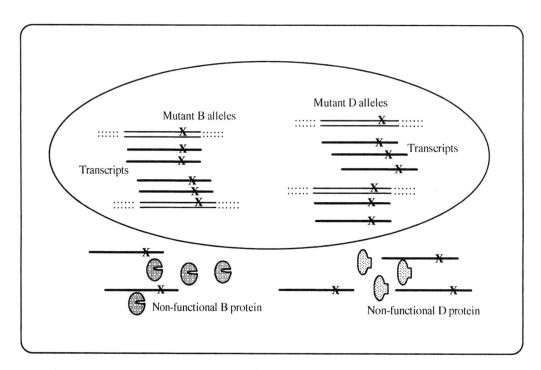

18. Sweet peas are able to produce two different pigments or colors. Just like mixing paints or crayons, these two colors can be blended to produce a third. Only those plants that inherited the ability to make both blue and red

pigments have purple flowers. Some plants, however, do not inherit this ability. Some have lost the ability to produce red and therefore have blue flowers, and conversely, some have lost the ability to produce blue and have red flowers. If both of these abilities are lost (or not inherited), the plant has white flowers.

Solving the problem

    a.   A plant unable to synthesize red pigment would be blue.

    b.   A plant unable to synthesize blue pigment would be red.

    c.   A plant unable to synthesize both blue and red pigments would be white.

19.   Protein function can be destroyed by a mutation that causes the substitution of a single amino acid, even though the protein has the same immunological properties. For example, enzymes require very specific amino acids in exact positions within their active site. A substitution of one of these key amino acids might have no effect on overall size and shape of the protein while completely destroying its enzymatic activity.

20. a.   The mothers had an excess of phenylalanine in their blood, and that excess was converted to phenylpyruvic acid. The phenylpyruvic acid was passed through the placenta into the fetal circulatory system, where it caused brain damage prior to birth.

    b.   The diet had no effect because the neurological damage has already happened *in utero* prior to birth.

    c.   A fetus with two mutant copies of the allele that causes PKU makes no functional enzyme. However, the mother of such a child is heterozygous and makes enough enzyme to block any brain damage; the excess phenylalanine in the fetal circulatory system enters the maternal circulatory system and is processed by the maternal gene product. After birth, which is when PKU damage occurs in a PKU child, dietary restrictions block a buildup of phenylalanine in the circulatory system until brain development is completed.

    The fetus of a mother with PKU is exposed to phenylpyruvic acid in its circulatory system during the time of major brain development. Therefore, brain damage occurs before birth, and no dietary restrictions after birth can repair that damage.

    d.   The obvious solution to the brain damage seen in the babies of PKU mothers is to return the mother to a restricted diet during pregnancy in order to block high levels of exposure to her child.

e. PKU is characterized as a rare recessive disorder. A child with PKU has two parents that carry a mutant allele for the metabolism of phenylalanine. When two individuals who are heterozygous for PKU have a child, the risk that the child will have PKU is 25 percent. A PKU child is unable to make a functional enzyme that converts phenylalanine to tyrosine. As a result, an excess level of phenylalanine is found in the blood, and the excess is detected as an increase in phenylpyruvic acid in both the blood and the urine. The excess phenylpyruvic acid blocks normal development of the brain, resulting in retardation.

21. DNA has often been called the "blueprint of life" but how does it actually compare to a blueprint used for house construction? Both are abstract representations of instructions for building three-dimensional forms and both require interpretation for their information to be useful. But real blueprints are two-dimensional renderings of the various views of the final structure drawn to scale. There is one-to-one correlation between the lines on the drawing and the real form. The information in the DNA is encoded in a linear array — a one-dimensional set of instructions which only becomes three-dimensional as the encoded linear array of amino acids fold into their many forms. Also included in the informational content of the DNA are all the directions required for its "house" to maintain and repair itself, respond to change, and replicate. Let's see a real blueprint do that!

22. Other than the human ability to synthesize and wear synthetic materials, living systems are "proteins or something that has been made by a protein."

23. The formula *genotype + environment = phenotype* is not quite accurate. While phenotype is a product of the genotype and the environment, there is also an inherent variation due to random noise. Given complete information regarding both genotype and the environment, it is still impossible to specify the phenotype completely, although a close approximation can be achieved.

2

Single Gene Inheritance

BASIC PROBLEMS

1. The human genome contains an estimated 20,000–25,000 genes located on 23 different chromosomes.

2. PFGE separates DNA molecules by size. When DNA is carefully isolated from *Neurospora* (which has seven different chromosomes) seven bands should be produced using this technique. Similarly, the pea has seven different chromosomes and will produce seven bands (homologous chromosomes will co-migrate as a single band).

3. There is a total of 4 m of DNA and nine chromosomes per haploid set. On average, each is $4/9$ m long. At metaphase, their average length is 13 μm, so the average packing ratio is 13×10^{-6} m : 4.4×10^{-1} m or roughly 1:34,000! This remarkable achievement is accomplished through the interaction of the DNA with proteins. At its most basic, eukaryotic DNA is associated with histones in units called nucleosomes and during mitosis, coils into a solenoid. As loops, it associates with and winds into a central core of nonhistone protein called the scaffold.

4. Because the DNA levels vary four-fold, the range covers cells that are haploid (gametes) to cells that are dividing (after DNA has replicated but prior to cell division). The following cells would fit the DNA measurements:

    x haploid cells
    $2x$ diploid cells in G_1 or cells after meiosis I but prior to meiosis II
    $4x$ diploid cells after S but prior to cell division

5. The key function of mitosis is to generate two daughter cells genetically identical to the original parent cell.

6.   Two key functions of meiosis are to halve the DNA content and to reshuffle the genetic content of the organism to generate genetic diversity among the progeny.

7.   Its pretty hard to beat several billions of years of evolution but it might be simpler if DNA did not replicate prior to meiosis. The same events responsible for halving the DNA and producing genetic diversity could be achieved in a single cell division if homologous chromosomes paired, recombined, randomly aligned during metaphase, and separated during anaphase, etc. However, you would lose the chance to check and repair DNA that replication allows.

8.   In large part, this question is asking, why sex? Parthogenesis (the ability to reproduce without fertilization — in essence, cloning) is not common among multicellular organisms. Parthenogenesis occurs in a some species of lizards and fishes, and several kinds of insects but it is tbe only means of reproduction in only a few of these species. In plants, about 400 species can reproduce asexually by a process called apomixis. These plants produce seeds without fertilization. However, the majority of plants and animals reproduce sexually. Sexual reproduction produces a wide variety of different offspring by forming new combinations of traits inherited from both the father and the mother. Despite the numerical advantages of asexual reproduction, most multicellular species that have adopted it as their only method of reproducing have become extinct. However, there is no agreed upon explanation of why the loss of sexual reproduction usually leads to early extinction or conversely, why sexual reproduction is associated with evolutionary success.

On the other hand, the immediate effects of such a scenario are obvious. All offspring will be genetically identical to their mothers, and males would be extinct within one generation.

9.   As cells divide mitotically, each chromosome consists of identical sister chromatids that are separated to form genetically identical daughter cells. Although the second division of meiosis appears to be a similar process, the "sister" chromatids are likely to be different. Recombination during earlier meiotic stages has swapped regions of DNA between sister and nonsister chromosomes such that the two daughter cells of this division typically are not genetically identical.

10.   The four stages of mitosis are: prophase, metaphase, anaphase, and telophase. The first letters, PMAT, can be remembered by a mnemonic such as: Playful Mice Analyze Twice.

The five stages of prophase I are: leptotene, zygotene, pachytene, diplotene, and diakinesis. The first letters, LZPDD, can be remembered by a mnemonic such as: Large Zoos Provide Dangerous Distractions.

11. Yes. It could work but certain DNA repair mechanisms (such as postreplication recombination repair) could not be invoked prior to cell division. There would be just two cells as products of this meiosis, rather than four.

12. The nucleus contains the genome and separates it from the cytoplasm. However, during cell division, the nuclear envelope dissociates (breaks down). It is the job of the microtubule-based spindle to actually separate the chromosomes (divide the genetic material) around which nuclei reform during telophase. In this sense, it can be viewed as a passive structure that is divided by the cell's cytoskeleton.

13. Yes. Half of our genetic makeup is derived from each parent, each parent's genetic makeup is derived half from each of their parents, etc.

14. Because the "half" inherited is very random, the chances of receiving exactly the same half is vanishingly small. Ignoring recombination and focusing just on which chromosomes are inherited from one parent (for example, the one they inherited from their father or the one from their mother?), there are 2^{23} = 8,388,608 possible combinations!

15.

	Mitosis	Meiosis
fern	sporophyte gametophyte	sporophyte (sporangium)
moss	sporophyte gametophyte	sporophyte (atheridium and archegonium)
plant	sporophyte gametophyte	sporophyte (anther and ovule)
pine tree	sporophyte gametophyte	sporophyte (pine cone)
mushroom	sporophyte gametophyte	sporophyte (ascus or basidium)
frog	somatic cells	gonads
butterfly	somatic cells	gonads
snail	somatic cells	gonads

16. This problem is tricky because the answers depend on how a cell is defined. In general, geneticists consider the transition from one cell to two cells to occur with the onset of anaphase in both mitosis and meiosis, even though cytoplasmic division occurs at a later stage.

    a.   46 chromosomes, each with two chromatids = 92 chromatids

    b.   46 chromosomes, each with two chromatids = 92 chromatids

    c.   46 physically separate chromosomes in each of two about-to-be-formed cells

    d.   23 chromosomes in each of two about-to-be-formed cells, each with two chromatids = 46 chromatids

    e.   23 chromosomes in each of two about-to-be-formed cells

17.   (5) chromosome pairing (synapsis)

18.   First, examine the crosses and the resulting genotypes of the endosperm:

Female	Male	Polar nuclei	Sperm	Endosperm
f'/f' (floury)	f''/f''	f' and f'	f''/f''	$f'/f'/f''$
f''/f'' (flinty)	f'/f'	f'' and f''	f'/f'	$f''/f''/f'$

As can be seen, the phenotype of the endosperm correlates to the predominant allele present.

19.   Mendel's first law states that alleles are in pairs and segregate during meiosis, and his second law states that genes assort independently during meiosis.

20.   Do a testcross (cross to a/a). If the fly was A/A, all the progeny will be phenotypically A; if the fly was A/a, half the progeny will be A and half will be a.

21.   **a.**   A diploid meiocyte that is heterozygous for one gene (for example, s^+/s where s is the allele that confers the small colony phenotype) will, after replication and segregation, give two meiotic products of genotype s^+ and two of s. If the random spores of many meiocytes are analyzed, you would expect to find about 50 percent normally sized colonies and 50 percent small colonies if the abnormal phenotype is the result of a mutation in a single gene. Thus, the actual results of 188 normally sized and 180 small sized colonies supports the hypothesis that the phenotype is the result of a mutation in a single gene.

    b. The following represents an ascus with four spores. The important detail is that two of the spores are s and two are s^+.

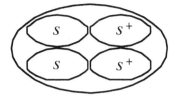

22. The progeny ratio is approximately 3:1, indicating classic heterozygous-by-heterozygous mating. Since Black (*B*) is dominant to white (*b*):

Parents: *B*/*b* × *B*/*b*

Progeny: 3 black:1 white (1 *B*/*B* : 2 *B*/*b* : 1 *b*/*b*)

23. a. You expect two *lys-5*⁺ (black) spores and two *lys-5* (white) spores.

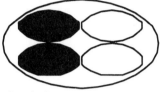

b. You expect all *lys-5* (white) spores.

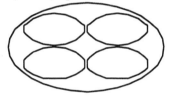

c. You expect all *lys-5*⁺ (black) spores.

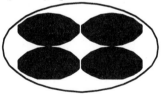

24. a. You do not expect the mutation to be recessive. This would be an example of a haploinsufficient gene since one copy of the wild-type allele does produce enough protein product for normal function.

b. An important assumption would be that having five of eight units of protein product would result in an observable phenotype. It also assumes that the regulation of the single wild-type allele is not affected. Finally, if the mutant allele was leaky rather than null, there might be sufficient protein function when heterozygous with a wild-type allele.

25. The simplest explanation is that the abnormal phenotype was not due to an genetic change. Perhaps the environment (edge of plate) was less favorable for growth. Since *Neurospora* is haploid and forms ascospores, isolating individual asci from a cross of the possible "mutant" to wild type and individually growing the spores should yield 50 percent wild-type and 50 percent "mutant" colonies. If all spores yield wild-type colonies the low density phenotype was not heritable.

26. Since half of the F_1 progeny are mutant, it suggests that the mutation that results in three cotyledons is dominant and the original mutant was heterozygous. Assuming C = the mutant allele and c = the wild-type allele, the cross becomes:

P $C/c \times c/c$

F_1 C/c three cotyledons
             c/c two cotyledons

27. a. The data for both crosses suggest that both A and B mutant plants are homozygous for recessive alleles. Both F_2 crosses give 3:1 ratios of normal to mutant progeny. For example, let A = normal and a = mutant, then

P $A/A \times a/a$
F_1 A/a
F_2 1 A/A phenotype: normal
             2 A/a phenotype: normal
             1 a/a phenotype: mutant (no trichomes).

  b. No. You do not know if the a and b mutations are in the same or different genes. If they are in the same gene then the F_1 will all be mutant. If they are in different genes, then the F_1 will all be wild type.

28. Each die has six sides, so the probability of any one side (number) is $^1/_6$.
To get specific red, green, and blue numbers involves "and" statements that are independent. So each independent probability is multiplied together.

  a. $(^1/_6)(^1/_6)(^1/_6) = (^1/_6)^3 = {}^1/_{216}$

  b. $(^1/_6)(^1/_6)(^1/_6) = (^1/_6)^3 = {}^1/_{216}$

  c. $(^1/_6)(^1/_6)(^1/_6) = (^1/_6)^3 = {}^1/_{216}$

  d.   To not roll any sixes is the same as getting anything but sixes:
       $(1 - ^1/_6)(1 - ^1/_6)(1 - ^1/_6) = (^5/_6)^3 = {}^{125}/_{216}.$

  e.   The easiest way to approach this problem is to consider each die separately. The first die thrown can be any number. Therefore, the probability for it is 1.

The second die can be any number except the number obtained on the first die. Therefore, the probability of not duplicating the first die is $1 - p$(first die duplicated) $= 1 - \frac{1}{6} = \frac{5}{6}$.

The third die can be any number except the numbers obtained on the first two dice. Therefore, the probability is $1 - p$(first two dice duplicated) $= 1 - \frac{2}{6} = \frac{2}{3}$.

Finally, the probability of all different dice is $(1)(\frac{5}{6})(\frac{2}{3}) = \frac{10}{18} = \frac{5}{9}$.

29. You are told that the disease being followed in this pedigree is very rare. If the allele that results in this disease is recessive, then the father would have to be homozygous and the mother would have to be heterozygous for this allele. On the other hand, if the trait is dominant, then all that is necessary to explain the pedigree is that the father is heterozygous for the allele that causes the disease. This is the better choice as it is more likely, given the rarity of the disease.

30.   **a.** By considering the pedigree (see below), you will discover that the cross in question is $T/t \times T/t$. Therefore, the probability of being a taster is $\frac{3}{4}$, and the probability of being a nontaster is $\frac{1}{4}$.

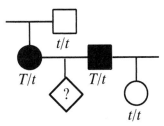

Also, the probability of having a boy equals the probability of having a girl equals $\frac{1}{2}$.

     (1) p(nontaster girl) $= p$(nontaster) $\times p$(girl) $= \frac{1}{4} \times \frac{1}{2} = \frac{1}{8}$

     (2) p(taster girl) $= p$(taster) $\times p$(girl) $= \frac{3}{4} \times \frac{1}{2} = \frac{3}{8}$

     (3) p(taster boy) $= p$(taster) $\times p$(boy) $= \frac{3}{4} \times \frac{1}{2} = \frac{3}{8}$

   b. p(taster for first two children) $= p$(taster for first child) $\times p$(taster for second child) $= \frac{3}{4} \times \frac{3}{4} = \frac{9}{16}$

31. *Unpacking the Problem*

1. Yes. The pedigree is given below.

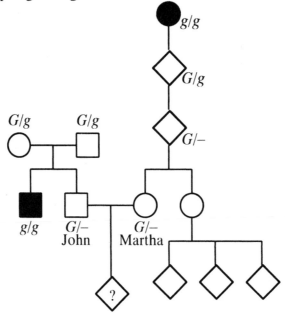

2. In order to state this problem as a Punnett square, you must first know the genotypes of John and Martha. The genotypes can be determined only through considering the pedigree. Even with the pedigree, however, the genotypes can be stated only as $G/-$ for both John and Martha.

   The probability that John is carrying the allele for galactosemia is $2/3$, rather than the $1/2$ that you might guess. To understand this, recall that John's parents must be heterozygous in order to have a child with the recessive disorder while still being normal themselves (the assumption of normalcy is based on the information given in the problem). John's parents were both G/g. A Punnett square for their mating would be:

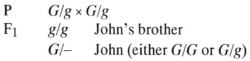

The cross is:

P	$G/g \times G/g$	
F_1	g/g	John's brother
	$G/-$	John (either G/G or G/g)

The expected ratio of the F_1 is 1 *G/G* : 2 *G/g* : 1 *g/g*. Because John does not have galactosemia (an assumption based on the information given in the problem), he can be either *G/G* or *G/g*, which occurs at a ratio of 1:2. Therefore, his probability of carrying the *g* allele is $^2/_3$.

The probability that Martha is carrying the *g* allele is based on the following chain of logic. Her great-grandmother had galactosemia, which means that she had to pass the allele to Martha's grandparent. Because the problem states nothing with regard to the grandparent's phenotype, it must be assumed that the grandparent was normal, or *G/g*. The probability that the grandparent passed it to Martha's parent is $^1/_2$. Next, the probability that Martha's parent passed the allele to Martha is also $^1/_2$, assuming that the parent actually has it. Therefore, the probability that Martha's parent has the allele and passed it to Martha is $^1/_2 \times ^1/_2$, or $^1/_4$.

In summary:

John $p(G/G) = ^1/_3$

           $p(G/g) = ^2/_3$

Martha $p(G/G) = ^3/_4$

           $p(G/g) = ^1/_4$

This information does not fit easily into a Punnett square.

3. While the above information could be put into a branch diagram, it does not easily fit into one and overcomplicates the problem, just as a Punnett square would.

4. The mating between John's parents illustrates Mendel's first law.

5. The scientific words in this problem are *galactosemia*, *autosomal*, and *recessive*.

Galactosemia is a metabolic disorder characterized by the absence of the enzyme galactose-1-phosphate uridyl transferase, which results in an accumulation of galactose. In the vast majority of cases, galactosemia results in an enlarged liver, jaundice, vomiting, anorexia, lethargy, and very early death if galactose is not omitted from the diet (initially, the child obtains galactose from milk).

Autosomal refers to genes that are on the autosomes.

Recessive means that in order for an allele to be expressed, it must be the only form of the gene present in the organism.

6. The major assumption is that if nothing is stated about a person's phenotype, the person is of normal phenotype. Another assumption that may be of value, but is not actually needed, is that all people marrying into these two families are normal and do not carry the allele for galactosemia.

7. The people not mentioned in the problem, but who must be considered, are John's parents and Martha's grandparent and parent descended from her affected great-grandmother.

8. The major statistical rule needed to solve the problem is the product rule (the "and" rule). It is used to calculate the cumulative probabilities described in part 2 of this unpacked solution (e.g., What is the probability that Martha's parent inherited the galactosemia allele AND passed that allele onto Martha AND Martha will pass that allele on to her child?).

9. Autosomal recessive disorders are assumed to be rare and to occur equally frequently in males and females. They are also assumed to be expressed if the person is homozygous for the recessive genotype.

10. Rareness leads to the assumption that people who marry into a family that is being studied do not carry the allele, which was assumed in entry (6) above.

11. The only certain genotypes in the pedigree are John's parents, John's brother, and Martha's great-grandmother and grandmother. All other individuals have uncertain genotypes.

12. John's family can be treated simply as a heterozygous-by-heterozygous cross, with John having a $2/3$ probability of being a carrier, while it is unknown if either of Martha's parents carry the allele. Therefore Martha's chance of being a carrier must be calculated as a series of probabilities.

13. The information regarding Martha's sister and her children turns out to be irrelevant to the problem.

14. The problem contains a number of assumptions that have not been necessary in problem solving until now.

15. Many scenarios are possible in response to this question.

Solution to the Problem

p(child has galactosemia) = p(John is G/g) × p(Martha is G/g) × p(both parents passed g to the child) = $(2/3)(1/4)(1/4) = 2/48 = 1/24$

32. Charlie, his mate, or both, obviously were not pure-breeding, because his F_2 progeny were of two phenotypes. Let A = black and white, and a = red and white. If both parents were heterozygous, then red and white would have been expected in the F_1 generation. Red and white were not observed in the F_1 generation, so only one of the parents was heterozygous. The cross is:

$$P \qquad A/a \; \times \; A/A$$
$$F_1 \qquad 1 \, A/a \; : \; 1 \, A/A$$

Two F_1 heterozygotes (A/a) when crossed would give 1 A/A (black and white) : 2 A/a (black and white) : 1 a/a (red and white). If the red and white F_2 progeny were from more than one mate of Charlie's, then the farmer acted correctly. However, if the F_2 progeny came only from one mate, the farmer may have acted too quickly.

33. Because the parents are heterozygous, both are A/a. Both twins could be albino or both twins could be normal (and = multiply, or = add). The probability of being normal ($A/-$) is $3/4$, and the probability of being albino (a/a) is $1/4$.

$$p(\text{both normal}) \; + p(\text{both albino})$$
$$p(\text{first normal}) \times p(\text{second normal}) + p(\text{first albino}) \times p(\text{second albino})$$
$$(3/4)(3/4) + (1/4)(1/4) = 9/16 + 1/16 = 5/8$$

34. The plants are approximately 3 blotched : 1 unblotched. This suggests that blotched is dominant to unblotched and that the original plant which was selfed was a heterozygote.

    a. Let A = blotched, a = unblotched.
        P A/a (blotched) $\times A/a$ (blotched)
        F_1 $1\,A/A : 2\,A/a : 1\,a/a$
              $3\,A/-$ (blotched) : 1 a/a (unblotched)

    b. All unblotched plants should be pure-breeding in a testcross with an unblotched plant (a/a), and one-third of the blotched plants should be pure-breeding.

35. In theory, it cannot be proved that an animal is not a carrier for a recessive allele. However, in an $A/- \times a/a$ cross, the more dominant-phenotype progeny produced, the less likely it is that the parent is A/a. In such a cross, half the progeny would be a/a and half would be A/a. With n dominant phenotype progeny, the probability that the parent is A/a is $(1/2)^n$. (DNA sequencing can be used to prove heterozygosity, but without sequence level information, the level of certainty is limited by sample size.)

36. The results suggest that winged ($A/-$) is dominant to wingless (a/a) (cross 2 gives a 3 : 1 ratio). If that is correct, the crosses become

		Number of progeny plants	
Pollination	Genotypes	Winged	Wingless
winged (selfed)	$A/A \times A/A$	91	1*
winged (selfed)	$A/a \times A/a$	90	30
wingless (selfed)	$a/a \times a/a$	4*	80
winged × wingless	$A/A \times a/a$	161	0
winged × wingless	$A/a \times a/a$	29	31
winged × wingless	$A/A \times a/a$	46	0
winged × winged	$A/A \times A/-$	44	0
winged × winged	$A/A \times A/-$	24	0

The five unusual plants are most likely due either to human error in classification or to contamination. Alternatively, they could result from environmental effects on development. For example, too little water may have prevented the seed pods from becoming winged even though they are genetically winged.

37. **a.** The disorder appears to be dominant because all affected individuals have an affected parent. If the trait was recessive, then I-1, II-2, III-1, and III-8 would all have to be carriers (heterozygous for the rare allele).

b. Assuming dominance, the genotypes are
I : $d/d, D/d$
II : $D/d, d/d, D/d, d/d$
III : $d/d, D/d, d/d, D/d, d/d, d/d, D/d, d/d$
IV : $D/d, d/d, D/d, d/d, d/d, d/d, d/d, D/d, d/d$

c. The mating is $D/d \times d/d$. The probability of an affected child (D/d) equals $1/2$, and the probability of an unaffected child (d/d) equals $1/2$. Therefore, the chance of having four unaffected children (since each is an independent event) is: $1/2 \times 1/2 \times 1/2 \times 1/2 = 1/16$.

38. **a.** *Pedigree 1*: The best answer is recessive, because two unaffected individuals had affected progeny. Also, the disorder skips generations and appears in a mating between two related individuals.

Pedigree 2: The best answer is dominant, because two affected parents have an unaffected child. Also, it appears in each generation, roughly half the progeny are affected, and all affected individuals have an affected parent.

Pedigree 3: The best answer is dominant, for many of the reasons stated for pedigree 2. Inbreeding, while present in the pedigree, does not allow an explanation of recessive because it cannot account for individuals in the second or third generations.

Pedigree 4: The best answer is recessive. Two unaffected individuals had affected progeny.

b. Genotypes of pedigree 1:
  Generation I: *A/–, a/a*
  Generation II: *A/a, A/a, A/a, A/–, A/–, A/a*
  Generation III: *A/a, A/a*
  Generation IV: *a/a*

Genotypes of pedigree 2:
  Generation I: *A/a, a/a, A/a, a/a*
  Generation II: *a/a, a/a, A/a, A/a, a/a, a/a, A/a, A/a, a/a*
  Generation III: *a/a, a/a, a/a, a/a, a/a, A/–, A/–, A/–, A/a, a/a*
  Generation IV. *a/a, a/a, a/a*

Genotypes of pedigree 3:
  Generation I: *A/–, a/a*
  Generation II: *A/a, a/a, a/a, A/a*
  Generation III: *a/a, A/a, a/a, a/a, A/a, a/a*
  Generation IV. *a/a, A/a, A/a, A/a, a/a, a/a*

Genotypes of pedigree 4:
  Generation I: *a/a, A/–, A/a, A/a*
  Generation II: *A/a, A/a, A/a, a/a, A/–, a/a, A/–, A/–, A/–, A/–, A/–*
  Generation III: *A/a, a/a, A/a, A/a, a/a, A/a*

39. a. The pedigree is

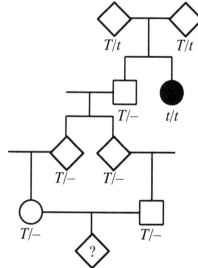

b. The probability that the child of the two first cousins will have Tay-Sachs disease is a function of three probabilities: p(the woman is T/t) $\times$ p(the man is T/t) $\times$ p(both donate t);

$$= (^2/_3)(^1/_2)(^1/_2) \times (^2/_3)(^1/_2)(^1/_2) \times ^1/_4 = ^1/_{144}$$

To understand the probabilities of the first two events, see the discussion for problem 31 part (2) of this chapter.

40. a. Autosomal recessive: affected individuals inherited the trait from unaffected parents and a daughter inherited the trait from an unaffected father.

b. Both parents must be heterozygous to have a $^1/_4$ chance of having an affected child. Parent 2 is heterozygous, since her father is homozygous for the recessive allele and parent 1 has a $^1/_2$ chance of being heterozygous, since his father is heterozygous because 1's paternal grandmother was affected. Overall, $1 \times ^1/_2 \times ^1/_4 = ^1/_8$.

41. a. Yes. It is inherited as an autosomal dominant trait.

b. Susan is highly unlikely to have Huntingtons disease. Her great-grandmother (individual II-2) is 75 years old and has yet to develop it, when nearly 100 percent of people carrying the allele will have developed the disease by that age. If her great-grandmother does not have it, Susan cannot inherit it.

Alan is somewhat more likely than Susan to develop Huntingtons disease. His grandfather (individual III-7) is only 50 years old, and approximately 20 percent of the people with the allele have yet to develop the disease by that age. Therefore it can be estimated that the grandfather has a 10 percent chance of being a carrier (50 percent chance he inherited the allele from his father $\times$ 20 percent chance he has not yet developed symptoms). If Alan's grandfather eventually develops Huntingtons disease, then there is a probability of 50 percent that Alan's father inherited it from him, and a probability of 50 percent that Alan received that allele from his father. Therefore, Alan has a $^1/_{10} \times ^1/_2 \times ^1/_2 = ^1/_{40}$ current probability of developing Huntingtons disease and $^1/_2 \times ^1/_2 = ^1/_4$ probability, if his grandfather eventually develops it.

42. a. Assuming the trait is rare, expect that all individuals marrying into the pedigree do not carry the disease-causing allele.

I : $P/P, p/p, p/p, P/P$
II : $P/P, P/p, P/p, P/p, P/p$
III : $P/P, P/-, P/-, P/P$
IV : $P/-, P/-$

b. For their child to have PKU, both A and B must be carriers and both must donate the recessive allele.

The probability that individual A has the PKU allele is derived from individual II-2. II-2 must be *P/p* since her father must be *p/p*. Therefore, the probability that II-2 passed the PKU allele to individual III-2 is $1/2$. If III-2 received the allele, the probability that he passed it to individual IV-1 (A) is $1/2$. Therefore, the probability that A is a carrier is $1/2 \times 1/2 = 1/4$.

The probability that individual B has the allele goes back to the mating of II-3 and II-4, both of whom are heterozygous. Their child, III-3, has a $2/3$ probability of having received the PKU allele and a probability of $1/2$ of passing it to IV-2 (B). Therefore, the probability that B has the PKU allele is $2/3 \times 1/2 = 1/3$.

If both parents are heterozygous, they have a $1/4$ chance of both passing the *p* allele to their child.
p(child has PKU) = p(A is *P/p*) × p(B is *P/p*) × p(both parents donate *p*)
$1/4$ × $1/3$ × $1/4$ = $1/48$

c. If the first child is normal, no additional information has been gained and the probability that the second child will have PKU is the same as the probability that the first child will have PKU, or $1/48$.

d. If the first child has PKU, both parents are heterozygous. The probability of having an affected child is now $1/4$, and the probability of having an unaffected child is $3/4$.

43. a. Sons inherit their X chromosome from their mother. The mother has earlobes, the son does not. If the allele for earlobes is dominant and the allele for lack of earlobes recessive, then the mother could be heterozygous for this trait and the gene could be X-linked.

b. It is not possible from the data given to decide which allele is dominant. If lack of earlobes is dominant, then the father would be heterozygous and the son would have a 50 percent chance of inheriting the dominant "lack-of-earlobes" allele. If lack of earlobes is recessive, then the trait could be autosomal or X-linked but in either case, the mother would be heterozygous.

44. a. Let *C* stand for the normal allele and *c* stand for the allele that causes cystic fibrosis.

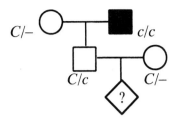

b. The man has a 100 percent probability of having the *c* allele. His wife, who is from the general population, has a $1/50$ chance of having the *c* allele. If both have the allele, then $1/4$ of their children will have cystic fibrosis. The probability that their first child will have cystic fibrosis is

$$p(\text{man has } c) \times p(\text{woman has } c) \times p(\text{both pass } c \text{ to the child})$$
$$1.0 \quad \times \quad 1/50 \quad \times \quad 1/4 = 1/200 = 0.005$$

c. If the first child does have cystic fibrosis, then the woman is a carrier of the *c* allele. Because both parents are *C/c*, the chance that the second child will be normal is the probability of a normal child in a heterozygous × heterozygous mating, or $3/4$.

45. The cross is *C/c* × *c/c* so there is a $1/2$ chance that a progeny would be black (*C/c*). Because each progeny's genotype is independent of the others, the chance that all 10 progeny are black is $(1/2)^{10}$.

46. P s^+/s^+ × s/Y
                 ↓

F₁ $1/2\ s^+/s$ normal female
       $1/2\ s^+/Y$ normal male

       s^+/s × s^+/Y
                 ↓

F₂ $1/4\ s^+/s^+$ normal female
       $1/4\ s^+/s$ normal female
       $1/4\ s^+/Y$ normal male
       $1/4\ s/Y$ small male

P s^+/s × s/Y
                 ↓

Progeny $1/4\ s^+/s$ normal female
         $1/4\ s/s$ small female
         $1/4\ s^+/Y$ normal male
         $1/4\ s/Y$ small male

47. Let H = hypophosphatemia and h = normal. The cross is $H/Y \times h/h$, yielding H/h (females) and h/Y (males). The answer is 0%.

48. a. You should draw pedigrees for this question.

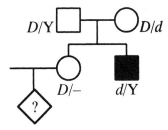

The "maternal grandmother" had to be a carrier, D/d. The probability that the woman inherited the d allele from her is $1/2$. The probability that she passes it to her child is $1/2$. The probability that the child is male is $1/2$. The total probability of the woman having an affected child is $1/2 \times 1/2 \times 1/2 = 1/8$.

b. The same pedigree as part (a) applies. The "maternal grandmother" had to be a carrier, D/d. The probability that your mother received the allele is $1/2$. The probability that your mother passed it to you is $1/2$. The total probability is $1/2 \times 1/2 = 1/4$.

c.

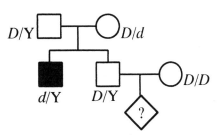

Because your father does not have the disease, you cannot inherit the allele from him. Therefore, the probability of inheriting an allele will be based on the chance that your mother is heterozygous. Since she is "unrelated" to the pedigree, assume that this is zero.

49. For the recently married man and woman to each have an uncle with alkaptonuria means that each may have one parent (the parent related to the uncle) that is heterozygous for the disease-causing allele. Specifically, this parent (and related uncle) must have had parents that were both heterozygous for alkaptonuria. Any child of parents that are both heterozygous for a recessive trait, but does not have that trait, has a $2/3$ chance of being heterozygous. (Remember, if both parents are heterozygous, we expect a 1 : 2 : 1 ratio of genotypes but once we know a person is not homozygous recessive, the only possibilities left are 1 (homozygous dominant) to 2 (heterozygous) or $2/3$ chance of being heterozygous.) So the man and woman each have a $2/3 \times 1/2 = 1/3$ of being carriers (heterozygous) and the chance of their having an affected child would be $1/3 \times 1/3 \times 1/4 = 1/36$.

50. **a.** Because none of the parents are affected, the disease must be recessive. Because the inheritance of this trait appears to be sex-specific, it is most likely X-linked. If it were autosomal, all three parents would have to be carriers and by chance, only sons and none of the daughters inherited the trait (which is quite unlikely).

    b.   I $A/Y, A/a, A/Y$

          II $A/Y, A/-, a/Y, A/-, A/Y, a/Y, a/Y, A/-, a/Y, A/-$

51. **a.** This question is similar to question 49, but this time the discussion begins with the grandparents rather than the parents. Again, given that a sibling is affected with a recessive disease, the related unaffected brother/sister will have a $2/3$ chance of being heterozygous. In this case, that is one of the grandparents of both the man and woman about to be married. Given this, the couple will both have a $2/3 \times 1/2 \times 1/2 = 1/6$ chance of being carriers (heterozygous) and the chance of their having an affected child will be $1/6 \times 1/6 \times 1/4 = 1/144$.

    b. If both parents are carriers, there is a $3/4$ chance a child will be normal and a $1/4$ chance a child will have cystic fibrosis. Each child is an independent event but since birth order is not considered, there are four ways to have the desired outcome. The child with cystic fibrosis may be the first, second, third, or fourth so assuming the first is affected, the specified outcome would be a $1/4 \times 3/4 \times 3/4 \times 3/4$ or a $27/256$ chance. Now taking into account the four possible birth orders and the chance that both parents are heterozygous, the chance of an exact 3:1 ratio becomes $4 \times 1/6 \times 1/6 \times 27/256 = 3/256$.

    c. In this case, knowing the first child has cystic fibrosis lets us now deduce that the parents must both be heterozygous. Given this, there is a $3/4$ chance than any future child will be normal. Since each is independent, the chance that their next three are normal is simply $3/4 \times 3/4 \times 3/4$ or $27/64$.

52. You should draw the pedigree before beginning.

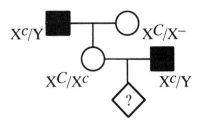

a. X^C/X^c, X^c/X^c

b. $p(\text{colorblind}) \times p(\text{male}) = (^1/_2)(^1/_2) = {}^1/_4$

c. The girls will be 1 normal (X^C/X^c) : 1 colorblind (X^c/X^c).

d. The cross is $X^C/X^c \times X^c/Y$, yielding 1 normal : 1 colorblind for both sexes.

53. **a.** This problem involves X-inactivation. Let B = black and b = *orange*.

Females	Males
X^B/X^B = black	X^B/Y = black
X^b/X^b = orange	X^b/Y = orange
X^B/X^b = calico	

b. P X^b/X^b (orange) $\times$ X^B/Y (black)
    F_1 X^B/X^b (calico female)
        X^b/Y (orange male)

c. Because the males are black or orange, the mother had to have been calico. Half the daughters are black, indicating that their father was black.

d. Males were orange or black, indicating that the mothers were calico. Orange females indicate that the father was orange.

e. Males were orange or black, indicating that the mothers were calico. Orange females indicate that the father was orange.

54. **a.** Recessive (unaffected parents have affected progeny) and X-linked (only assumption is that the grandmother, I-2, is a carrier). If autosomal, then I-1, I-2, and II-6 would all have to be carriers.

b. Generation I: X^A/Y, X^A/X^a
    Generation II: X^A/X^A, X^a/Y, X^AY, X^A/X^-, X^A/X^a, X^A/Y
    Generation III: X^A/X^A, X^A/Y, X^A/X^a, X^A/X^a, X^A/Y, X^AX^A, X^a/Y, X^A/Y, X^A/X^-

c. Because it is stated that the trait is rare, the assumption is that no one marrying into the pedigree carries the recessive allele. Therefore, the first couple has no chance of an affected child because the son received a Y chromosome from his father. The second couple has a 50 percent chance of having affected sons and no chance of having affected daughters. The third couple has no chance of having an affected child, but all of their daughters will be carriers.

55. Remember, the endosperm is formed from two polar nuclei (which are genetically identical) and one sperm nucleus.

Female	Male	Polar nuclei	Sperm	Endosperm
s/s	*S/S*	*s* and *s*	*S*	*S/s/s*
S/S	*s/s*	*S* and *S*	*s*	*S/S/s*
S/s	*S/s*	$^1/_2$ *S* and *S*	$^1/_2$ *S*	$^1/_4$ *S/S/S*
				$^1/_4$ *S/S/s*
		$^1/_2$ *s* and *s*	$^1/_2$ *s*	$^1/_4$ *S/s/s*
				$^1/_4$ *s/s/s*

56. To use the chi-square test, first state the hypothesis being tested and the expected results. In this case, the hypothesis is that the phenotypic difference is due to two alleles of one gene. The expected results would be 300 purple and 100 blue (or an expected 3:1 ratio in the F_2), The formula to use is

$$\chi^2 = \sum \text{(observed-expected)}^2/\text{expected}$$

Class	Observed (*O*)	Expected (*E*)	$(O–E)^2$	$(O–E)^2/E$
purple	320	300	400	1.33
blue	80	100	400	4.00

Total = χ^2 = 5.33

This χ^2 value must now be looked up in a table (Table 2-2 in the companion text) which will give you the probability that these data fit the stated hypothesis. But first, you must also determine the degrees of freedom (df) or the number of independent variables in the data. Generally, the degrees of freedom can be calculated as the number of phenotypes in the problem minus one. In this example, there are two phenotypes (purple and blue) and therefore, one degree of freedom. Using the table, you will find that the data have a *p* value that is between 0.025 and 0.01. The *p* value is the probability that the deviation of the observed from that expected is due to chance alone. The relative standard commonly used in biological research for rejecting a hypothesis is $p < 0.05$. In this case, the data do not support the hypothesis.

57. **a.** Galactosemia pedigree

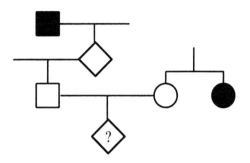

b. Both parents must be heterozygous for this child to have a $1/4$ chance of inheriting the disease. Since the mother's sister is affected with galactosemia, their parents must have both been heterozygous. Since the mother does not have the trait, there is a $2/3$ chance that she is a carrier (heterozygous). One of the father's parents must be a carrier since his grandfather had the recessive trait. Thus, the father had a $1/2$ chance of inheriting the allele from that parent. Since these are all independent events, the child's risk is

$$1/4 \times 2/3 \times 1/2 = 1/12$$

c. If the child has galactosemia, both parents must be carriers and thus those probabilities become 100 percent. Now all future children have a $1/4$ chance of inheriting the disease.

CHALLENGING PROBLEMS

58. The probability of obtaining y/y ; r/r from this cross is $1/16$, and the probability of not obtaining this is $15/16$. Since only one plant is needed, the probability of not getting this genotype in n trials is $(15/16)^n$. Because the probability of failure must be no greater than 5 percent:

$$(15/16)^n < 0.05$$
$$n > 46.42, \text{ or } 47 \text{ plants}$$

59. **a.** In order to draw this pedigree, you should realize that if an individual's status is not mentioned, then there is no way to assign a genotype to that person. The parents of the boy in question had a phenotype (and genotype) that differed from his. Therefore, both parents were heterozygous and the boy, who is a non-roller, homozygous recessive. Let R stand for the ability to roll the tongue and r stand for the inability to roll the tongue. The pedigree becomes:

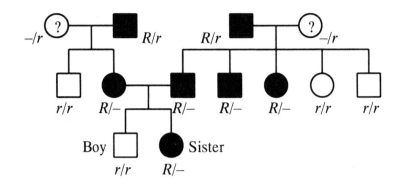

b. Assuming the twins are identical, there might be either an environmental component to the expression of that gene or developmental noise (see Chapter 1). Another possibility is that the *R* allele is not fully penetrant and that some genotypic "rollers" do not express the phenotype.

60. a. The inheritance pattern for red hair suggested by this pedigree is recessive since most red-haired individuals are from parents without this trait.

   b. In most populations, the allele appears to be somewhat rare.

61. *Taster by taster cross:* Tasters can be either *T/T* or *T/t*, and the genotypic status cannot be determined until a large number of progeny are observed. A failure to obtain a 3 : 1 ratio in the matings of two tasters would be expected because there are three types of matings:

Mating	Genotypes	Phenotypes
T/T × *T/T*	all *T/T*	all tasters
T/T × *T/t*	1 *T/T* : 1 *T/t*	all tasters
T/t × *T/t*	1 *T/T* : 2 *T/t* : 1 *t/t*	3 tasters : 1 non-tasters

Taster by non-taster cross: There are two types of matings that resulted in the observed progeny:

Mating	Genotypes	Phenotypes
T/T × *t/t*	all *T/t*	all tasters
T/t × *t/t*	1 *T/t* : 1 *t/t*	1 tasters : 1 non-tasters

Again, the failure to obtain either a 1:0 ratio or a 1:1 ratio would be expected because of the two mating types.

Non-taster by non-taster cross: There is only one mating that is non-taster by non-taster (*t/t* × *t/t*), and 100 percent of the progeny would be expected to be non-tasters. Of 223 children, five were classified as tasters. Some could be the result of mutation (unlikely), some could be the result of misclassification (likely), some could be the result of a second gene that affects the expression of the gene in question (possible), some could be the result of developmental noise (possible), and some could be due to illegitimacy (possible).

62. If the historical record is accurate, the data suggest Y linkage. Another explanation is an autosomal gene that is dominant in males and recessive in females. This has been observed for other genes in both humans and other species.

63. The different sex-specific phenotypes found in the F_1 indicate sex-linkage — the females inherit the trait of their fathers. The first cross also indicates that the wild-type large spots is dominant over the lacticolor small spots. Let A = wild type and a = lacticolor.

Cross 1: If the male is assumed to be the hemizygous sex, then it soon becomes clear that the predictions do not match what was observed:

P a/a female × A/Y male

F_1 A/a wild-type females
        a/Y lacticolor males

Therefore, assume that the female is the hemizygous sex. Let Z stand for the sex-determining chromosome in females. The cross becomes:

P a/Z female × A/A male

F_1 A/a wild-type male
        A/Z wild-type female

F_2 $1/4$ A/Z wild-type females
        $1/2$ $A/-$ wild-type males
        $1/4$ a/Z lacticolor females

Cross 2:

P A/Z female × a/a male

F_1 a/Z lacticolor females
        A/a wild-type males
F_2 $1/4$ A/Z wild-type females
        $1/4$ A/a wild-type males
        $1/4$ a/Z lacticolor females
        $1/4$ a/a lacticolor males

64. Note that only males are affected and that in all but one case, the trait can be traced through the female side. However, there is one example of an affected male having affected sons. If the trait is X-linked, this male's wife must be a carrier. Depending on how rare this trait is in the general population, this suggests that the disorder is caused by an autosomal dominant with expression limited to males.

65. Because the disorder is X-linked recessive, the affected male had to have received the allele, *a*, from the female common ancestor in the first generation. The probability that the affected man's wife also carries the *a* allele is the probability that she also received it from the female common ancestor. That probability is $1/8$.

The probability that the couple will have an affected boy is
$$p(\text{father donates Y}) \times p(\text{the mother has } a) \times p(\text{mother donates } a)$$
$$1/2 \times \quad 1/8 \times \quad 1/2 = 1/32$$

The probability that the couple will have an affected girl is:
$$p(\text{father donates } X^a) \times p(\text{the mother has } a) \times p(\text{mother donates } a)$$
$$1/2 \times \quad 1/8 \times \quad 1/2 = 1/32$$

The probability of normal children:
$$= 1 - p(\text{affected children})$$
$$= 1 - p(\text{affected male}) - p(\text{affected female})$$
$$= 1 - 1/32 - 1/32 = 30/32 = 15/16$$

Half the normal children will be boys, with a probability of $15/32$, and half will be girls, with a probability of $15/32$.

66. a. The complete absence of male offspring is the unusual aspect of this pedigree. In addition, all progeny that mate carry the trait for lack of male offspring. If the male lethality factor were nuclear, the male parent would be expected to alter this pattern. Therefore, cytoplasmic inheritance is suggested.

b. If all females resulted from chance alone, then the probability of this result is $(1/2)^n$, where n = the number of female births. In this case n is 72. Chance is an unlikely explanation for the observations.

The observations can be explained by cytoplasmic factors by assuming that the proposed mutation in mitochondria is lethal only in males.

A modified form of Mendelian inheritance, an autosomal dominant, sex-limited lethal trait, might also explain these data but it is an an unlikely answer, due to the probability arguments above.

Tips on Problem Solving

Ratios: a 1:2:1 (or 3:1) ratio indicates that one gene is involved (see Problem 22). A 9:3:3:1 ratio, or some modification of it, indicates that two genes are involved (see Chapter 3). A testcross results in a 1:1 ratio if the organism being tested is heterozygous and a 1:0 ratio if it is homozygous (see Problem 20).

Pedigrees: normal parents have affected offspring in recessive disorders. Normal parents have normal offspring and affected parents have affected and/or normal offspring in dominant disorders). If phenotypically identical parents produce progeny with two phenotypes, the parents were both heterozygous (see Problem 38).

Probability: when dealing with two or more independently assorting genes, consider each gene separately.

X-linked or autosomal: if the male phenotype is different from the female phenotype, X linkage is involved for the allele carried by the female (see Problems 47, 48).

Inheritance patterns: there are only seven possible inheritance patterns for a gene. Usually only numbers 1–4 will be encountered:

1. Autosomal dominant
2. Autosomal recessive
3. X-linked dominant
4. X-linked recessive
5. Autosomal with expression limited to one sex
6. Y-linked
7. X- and Y-linked (pseudoautosomal)

A Systematic Approach to Problem Solving

Now that you have struggled with a number of genetics problems, it may be worthwhile to make some generalizations about problem solving beyond what has been presented for each chapter so far.

The first task always is to determine exactly what information has been presented and what is being asked. Frequently, it is necessary to rewrite the problem or to symbolize the presented information in some way.

The second task is to formulate and test hypotheses. If the results generated by a hypothesis contradict some aspect of the problem, then the hypothesis is rejected. If the hypothesis generates data compatible with the problem, then it is retained.

A systematic approach is the only safe approach in working genetics problems. Shortcuts in thought processes usually lead to an incorrect answer. Consider the following two types of problems.

1.   When analyzing pedigrees, there are usually only four possibilities (hypotheses) to be considered: autosomal dominant, autosomal recessive, X-linked dominant, and X-linked recessive. The criteria for each should be checked against the data. Additional factors that should be kept in mind are epistasis, penetrance, expressivity, age of onset, incorrect diagnosis in earlier generations, adultery, adoptions that are not mentioned, and inaccurate information in general. All these factors can be expected in real life, although few will be encountered in the problems presented here.

2.   When studying matings, frequently the first task is to decide whether you are dealing with one gene, two genes, or more than two genes (hypotheses). The location of the gene or genes may or may not be important. If location is important, then there are two hypotheses: autosomal and X-linked. If there are two or more genes, then you may have to decide on linkage relationships between them. There are two hypotheses: unlinked and linked.

If ratios are presented, then 1:2:1 (or some modification signaling dominance) indicates one gene, 9:3:3:1 (or some modification reflecting epistasis) indicates two genes, and 27:9:9:9:3:3:3:1 (or some modification signaling epistasis) indicates three genes. If ratios are presented that bear no relationship to the above, such as 35:35:15:15, then you are dealing with two linked genes (see Chapter 4 for a discussion of linkage).

If phenotypes rather than ratios are emphasized in the problem, then a cross of two mutants that results in wild type indicates the involvement of two genes rather than alleles of the same gene. Both mutants are recessive to wild type. A correlation of sex with phenotype indicates X linkage for the gene mutant in the female parent, while a lack of correlation indicates autosomal location.

If the problem involves X linkage, frequently the only way to solve it is to focus on the male progeny.

Once you determine the number of genes being followed and their location, the problem essentially solves itself if you make a systematic listing of genotype and phenotype.

Sometimes, the final portion of a problem will give additional information that requires you to adjust all the work that you have done up to that point. As an example, in Problem 64 of Chapter 6, crosses 1 – 4 lead you to assume that you are working with one gene. In cross 5, data incompatible with this assumption are presented. Your initial assumption of one gene is correct for the information given in the first four crosses; it is not a mistake. Other than a lack of systematic thought, the greatest mistake that a student can make is to label a rejected hypothesis an error. This decreases self-confidence and increases anxiety, with the result that real mistakes will likely follow. The beginner needs to keep in mind that science progresses by the rejection of hypotheses. When a hypothesis is rejected, something concrete is known: the proposed hypothesis does not explain the results. An unrejected hypothesis may be right or it may be wrong, and there is no way to know without further experimentation.

A very generalized procedure for problem solving would look like this:

1. Determine what information is being presented and what is being asked.

2. Formulate all possible hypotheses.

3. Check the consequences of each hypothesis against the data (the given information).
    Reject all hypotheses that are incompatible with the data.
    Retain all hypotheses that are compatible with the data.
    If no hypothesis is compatible with the data, return to step 1.

3

Independent Assortment of Genes

BASIC PROBLEMS

1. **a.** The expected phenotypic ratio from the self cross of A/a ; B/b is

9	$A/—$; $B/—$
3	$A/—$; b/b
3	a/a ; $B/—$
1	a/a ; b/b

    b. The expected genotypic ratio from the self cross of A/a ; B/b is

1	A/A ; B/B
2	A/A ; B/b
1	A/A ; b/b
2	A/a ; B/B
4	A/a ; B/b
2	A/a ; b/b
1	a/a ; B/B
2	a/a ; B/b
1	a/a ; b/b

    c. and d. The expected phenotypic and genotypic ratios from the test cross of A/a ; B/b is

1	A/a ; B/b
1	A/a ; b/b
1	a/a ; B/b
1	a/a ; b/b

2. The resulting cells will have the identical genotype as the original cell: A/a ; B/b.

3. The general formula for the number of different male/female centromeric combinations possible is 2^n, where n = number of different chromosome pairs. In this case, $2^5 = 32$.

4. Because the DNA levels vary six-fold, the range covers cells that are haploid (spores or cells of the gametophyte stage) to cells that are triploid (the endosperm) and dividing (after DNA has replicated but prior to cell division). The following cells would fit the DNA measurements:

0.7	haploid cells
1.4	diploid cells in G_1 or haploid cells after S but prior to cell division
2.1	triploid cells of the endosperm
2.8	diploid cells after S but prior to cell division
4.2	triploid cells after S but prior to cell division

5.

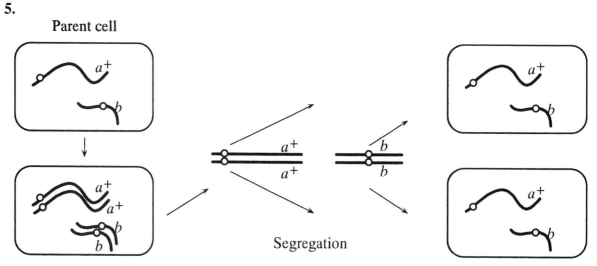

Parent cell

Chromosome duplication

Segregation

Daughter cells

6. a. A sporophyte of A/a ; B/b genotype will produce gametophytes in the following proportions:

$$^1/_4\,A\,;B$$
$$^1/_4\,A\,;b$$
$$^1/_4\,a\,;B$$
$$^1/_4\,a\,;b$$

b. Random fertilization of the spores from the above gametophytes can occur $4 \times 4 = 16$ possible ways. Four of these combinations (A ; $B \times a$; b, a ; $b \times A$; B, A ; $b \times a$; B, a ; $B \times A$; b) will result in the desired A/a ; B/b sporophyte genotype. Therefore $^1/_4$ of next generation should be of this genotype.

7. Mitosis produces cells with the same starting genotype: A/a ; B/b ; C/c.

8.   P ad^- ; a × ad^+ ; α

Transient diploid ad^+/ad^- ; a/α

    F_1 $^1/_4$ ad^+ ; a, white
                 $^1/_4$ ad^- ; a, purple
                 $^1/_4$ ad^+ ; α, white
                 $^1/_4$ ad^- ; α, purple

9. The cross is female X^d/X^d ; p/p × male X^D/Y ; P/P where P = dominant allele for pink and d = recessive allele for dwarf.

    F_1 $^1/_2$ X^D/X^d : P/p (pink female)
               $^1/_2$ X^d/Y ; P/p (dwarf, pink male)

    F_2 $^1/_{16}$ X^D/X^d : P/P (pink female)
               $^1/_8$ X^D/X^d : P/p (pink female)
               $^1/_{16}$ X^D/X^d : p/p (wild type female)
               $^1/_{16}$ X^d/X^d : P/P (dwarf, pink female)
               $^1/_8$ X^d/X^d : P/p (dwarf, pink female)
               $^1/_{16}$ X^d/X^d : p/p (dwarf female)
               $^1/_{16}$ X^D/Y ; P/P (pink male)
               $^1/_8$ X^D/Y ; P/p (pink male)
               $^1/_{16}$ X^D/Y ; p/p (wild type male)
               $^1/_{16}$ X^d/Y ; P/P (dwarf, pink male)
               $^1/_8$ X^d/Y ; P/p (dwarf, pink male)
               $^1/_{16}$ X^d/Y ; p/p (dwarf male)

10. His children will have to inherit the satellite-containing 4 (probability = $^1/_2$), the abnormally-staining 7 (probability = $^1/_2$), and the Y chromosome (probability = $^1/_2$). To get all three, the probability is $(^1/_2)(^1/_2)(^1/_2) = ^1/_8$.

11. The parental set of centromeres can match either parent, which means there are two ways to satisfy the problem. For any one pair, the probability of a centromere from one parent going into a specific gamete is $^1/_2$. For n pairs, the probability of all the centromeres being from one parent is $(^1/_2)^n$. Therefore, the total probability of having a haploid complement of centromeres from either parent is $2(^1/_2)^n = (^1/_2)^{n-1}$.

12.   Dear Monk Mendel:

I have recently read your most engrossing manuscript detailing the results of your most wise experiments with garden peas. I salute both your curiosity and your ingenuity in conducting said experiments, thereby opening up for scientific exploration an entire new area of our Maker's universe. Dear Sir, your findings are extraordinary!

While I do not pretend to compare myself to you in any fashion, I beg to bring to your attention certain findings I have made with the aid of that most fascinating and revealing instrument, the microscope. I have been turning my attention to the smallest of worlds with an instrument that I myself have built, and I have noticed some structures that may parallel in behavior the factors which you have postulated in the pea.

I have worked with grasshoppers, however, not your garden peas. Although you are a man of the cloth, you are also a man of science, and I pray that you will not be offended when I state that I have specifically studied the reproductive organs of male grasshoppers. Indeed, I did not limit myself to studying the organs themselves; instead, I also studied the smaller units that make up the male organs and have beheld structures most amazing within them.

These structures are contained within numerous small bags within the male organs. Each bag has a number of these structures, which are long and threadlike at some times and short and compact at other times. They come together in the middle of a bag, and then they appear to divide equally. Shortly thereafter, the bag itself divides, and what looks like half of the threadlike structures goes into each new bag. Could it be, Sir, that these threadlike structures are the very same as your factors? I know, of course, that garden peas do not have male organs in the same way that grasshoppers do, but it seems to me that you found it necessary to emasculate the garden peas in order to do some crosses, so I do not think it too far-fetched to postulate a similarity between grasshoppers and garden peas in this respect.

Pray, Sir, do not laugh at me and dismiss my thoughts on this subject even though I have neither your excellent training nor your astounding wisdom in the Sciences. I remain your humble servant to eternity!

13.   The hypothesis is that the organism being tested is a heterozygote and that the A/a and a/a progeny are of equal viability. The expected values would be that phenotypes occur with equal frequency. There are two genotypes in each case, so there is one degree of freedom.

$$\chi^2 = \sum (\text{observed-expected})^2/\text{expected}$$

a. $\chi^2 = [(120{-}110)^2 + (100{-}110)^2]/110$
$= 1.818; p > 0.10$, nonsignificant; hypothesis cannot be rejected

b. $\chi^2 = [(5000-5200)^2 + (5400-5200)^2]/5200$
$= 15.385; p < 0.005$, significant; hypothesis must be rejected

c. $\chi^2 = [(500-520)^2 + (540-520)^2]/520$
$= 1.538; p > 0.10$, nonsignificant; hypothesis cannot be rejected

d. $\chi^2 = [(50-52)^2 + (54-52)^2]/52$
$= 0.154; p > 0.50$, nonsignificant; hypothesis cannot be rejected

14. a. This is simply a matter of counting genotypes; there are nine genotypes in the Punnett square. Alternatively, you know there are three genotypes possible per gene, for example R/R, R/r, and r/r, and since both genes assort independently, there are $3 \times 3 = 9$ total genotypes.

b. Again, simply count. The genotypes are

1 R/R ; Y/Y	1 r/r ; Y/Y	1 R/R ; y/y	1 r/r ; y/y
2 R/r ; Y/Y	2 r/r ; Y/y	2 R/r ; y/y	
2 R/R ; Y/y			
4 R/r ; Y/y			

c. To find a formula for the number of genotypes, first consider the following:

Number of genes	Number of genotypes	Number of phenotypes
1	$3 = 3^1$	$2 = 2^1$
2	$9 = 3^2$	$4 = 2^2$
3	$27 = 3^3$	$8 = 2^3$

Note that the number of genotypes is 3 raised to some power in each case. In other words, a general formula for the number of genotypes is 3^n, where n equals the number of genes.

For allelic relationships that show complete dominance, the number of phenotypes is 2 raised to some power. The general formula for the number of phenotypes observed is 2^n, where n equals the number of genes.

d. The round, yellow phenotype is $R/-$; $Y/-$. Two ways to determine the exact genotype of a specific plant are through selfing or conducting a testcross.

With selfing, complete heterozygosity will yield a 9:3:3:1 phenotypic ratio. Homozygosity at one locus will yield a 3:1 phenotypic ratio, while homozygosity at both loci will yield only one phenotypic class.

With a testcross, complete heterozygosity will yield a 1:1:1:1 phenotypic ratio. Homozygosity at one locus will yield a 1:1 phenotypic ratio, while homozygosity at both loci will yield only one phenotypic class.

15.   Assuming independent assortment and simple dominant/recessive relationships of all genes, the number of genotypic classes expected from selfing a plant heterozygous for n gene pairs is 3^n and the number of phenotypic classes expected in 2^n.

16. a. The data for both crosses suggest that both A and B mutant plants are homozygous for a recessive allele. Both F_2 crosses give 3:1 normal to mutant ratios of progeny. For example, let A = normal and a = mutant, then

P $A/A \times a/a$
F_1 A/a
F_2 1 A/A phenotype: normal
          2 A/a phenotype: normal
          1 a/a phenotype: mutant (no trichomes).

b. The cross is $A/A\,;\,b/b \times a/a\,;\,B/B$ to give the F_1 of $A/a\,;\,B/b$. This is then test crossed (crossed to $a/a\,;\,b/b$) to give

$^1/_4$ $A/a\,;\,B/b$ (normal)
$^1/_4$ $A/a\,;\,b/b$ (no trichomes)
$^1/_4$ $a/a\,;\,B/b$ (no trichomes)
$^1/_4$ $a/a\,;\,b/b$ (no trichomes)

or 1 normal : 3 no trichomes.

17. a. $C/c\,;\,S/s \times C/c\,;\,S/s$ There are 3 short:1 long, and 3 dark:1 albino. Therefore, each gene is heterozygous in the parents.

b. $C/C\,;\,S/s \times C/-\,;\,s/s$ There are no albino, and there are 1 long:1 short indicating a testcross for this trait.

c. $C/c\,;\,S/S \times c/c\,;\,S/-$ There are no long, and there are 1 dark:1 albino.

d. $c/c\,;\,S/s \times c/c\,;\,S/s$ All are albino, and there are 3 short:1 long.

e. $C/c\,;\,s/s \times C/c\,;\,s/s$ All are long, and there are 3 dark:1 albino.

f. $C/C\,;\,S/s \times C/-\,;\,S/s$ There are no albino, and there are 3 short: 1 long.

g. $C/c\,;\,S/s \times C/c\,;\,s/s$ There are 3 dark:1 albino, and 1 short:1 long.

18. a. and b. Cross 2 indicates that purple (G) is dominant to green (g), and cross 1 indicates cut (P) is dominant to potato (p).

Cross 1 : $G/g\,;\,P/p \times g/g\,;\,P/p$ There are 3 cut : 1 potato, and 1 purple : 1 green.

Cross 2 : $G/g\,;\,P/p \times G/g\,;\,p/p$ There are 3 purple : 1 green, and 1 cut : 1 potato.

Cross 3: G/G ; P/p × g/g ; P/p There are no green, and there are
3 cut:1 potato.

Cross 4: G/g ; P/P × g/g ; p/p There are no potato, and there are
1 purple:1 green.

Cross 5: G/g ; p/p × g/g ; P/p There are 1 cut:1 potato, and there are
1 purple:1 green.

19. **a.** From cross 6, Bent (B) is dominant to normal (b). Both parents are "bent," yet some progeny are "normal."

   b. From cross 1, it is X-linked. The trait is inherited in a sex-specific manner — all sons have the mother's phenotype.

   c. In the following table, the Y chromosome is stated; the X is implied.

| | Parents | | Progeny | |
Cross	Female	Male	Female	Male
1	b/b	B/Y	B/b	b/Y
2	B/b	b/Y	$B/b, b/b$	$B/Y, b/Y$
3	B/B	b/Y	B/b	B/Y
4	b/b	b/Y	b/b	b/Y
5	B/B	B/Y	B/B	B/Y
6	B/b	B/Y	$B/B, B/b$	$B/Y, b/Y$

20. ***Unpacking the Problem***

   1. *Normal* is used to mean wild type, or red eye color and long wings.

   2. Both *line* and *strain* are used to denote pure-breeding fly stocks, and the words are interchangeable.

   3. Your choice.

   4. Three characters are being followed: eye color, wing length, and sex.

   5. For eye color, there are two phenotypes: red and brown. For wing length, there are two phenotypes: long and short. For sex, there are two phenotypes: male and female.

   6. The F_1 females designated normal have red eyes and long wings.

   7. The F_1 males that are called short-winged have red eyes and short wings.

   8. The F_2 ratio is

$3/8$ red eyes, long wings

$3/8$ red eyes, short wings

$1/8$ brown eyes, long wings

$1/8$ brown eyes, short wings

9. Because there is not the expected 9:3:3:1 ratio, one of the factors that distorts the expected dihybrid ratio must be present. Such factors can be sex linkage, epistasis, genes on the same chromosome, environmental effect, reduced penetrance, or a lack of complete dominance in one or both genes.

10. With sex linkage, traits are inherited in a sex-specific way. With autosomal inheritance, males and females have the same probabilities of inheriting the trait.

11. The F_2 does not indicate sex-specific inheritance.

12. The F_1 data does show sex-specific inheritance—all males are short-winged, like their mothers, while all females are normal-winged, like their fathers.

13. The F_1 suggests that long is dominant to short and red is dominant to brown. The F_2 data show a 3 red : 1 brown ratio indicating the dominance of red but a 1 : 1 long : short ratio indicative of a test cross. Without the F_1 data, it is not possible to determine which form of the wing character is dominant.

14. If Mendelian notation is used, then the red and long alleles need to be designated with uppercase letters, for example R and L, while the brown (r) and short (l) alleles need to be designated with lowercase letters. If *Drosophila* notation is used, then the brown allele may be designated with a lowercase b and the wild-type (red) allele with a b^+; the short wing-length gene with a s and the wild-type (long) allele with an s^+. (Genes are often named after their mutant phenotype.)

15. To deduce the inheritance of these phenotypes means to provide all genotypes for all animals in the three generations discussed and account for the ratios observed.

Solution to the Problem

Start this problem by writing the crosses and results so that all the details are clear.

P	brown, short female × red, long male
F_1	red, long females
	red, short males

These results tell you that red-eyed is dominant to brown-eyed and since both females and males are red-eyed, that this gene is autosomal. Since males differ from females in their genotype with regard to wing length, this trait is sex-linked. Knowing that *Drosophila* females are XX and males are XY, the long-winged females tell us that long is dominant to short and that the gene is X-linked. Let B = red, b = brown, S = long, and s = short. The cross can be rewritten as follows:

P b/b ; s/s × B/B ; S/Y

F$_1$ $^1/_2$ B/b ; S/s females
           $^1/_2$ B/b ; s/Y males

F$_2$ $^1/_{16}$ B/B ; S/s red, long, female
           $^1/_{16}$ B/B ; s/s red, short, female
           $^1/_8$ B/b ; S/s red, long, female
           $^1/_8$ B/b ; s/s red, short, female
           $^1/_{16}$ b/b ; S/s brown, long, female
           $^1/_{16}$ b/b ; s/s brown, short, female
           $^1/_{16}$ B/B ; S/Y red, long, male
           $^1/_{16}$ B/B ; s/Y red, short, male
           $^1/_8$ B/b ; S/Y red, long, male
           $^1/_8$ B/b ; s/Y red, short, male
           $^1/_{16}$ b/b ; S/Y brown, long, male
           $^1/_{16}$ b/b ; s/Y brown, short, male

The final phenotypic ratio is

$^3/_8$ red, long
$^3/_8$ red, short
$^1/_8$ brown, long
$^1/_8$ brown, short

with equal numbers of males and females in all classes.

21. a. Because photosynthesis is affected and the plants are yellow, rather than green, it is likely that the chloroplasts are defective. If the defect maps to the DNA of the chloroplast, the trait will be maternally inherited. This fits the data that all progeny have the phenotype of the female parent and not the phenotype of the male (pollen-donor) parent.

b. If the defect maps to the DNA of the chloroplast, the trait will be maternally inherited. Use pollen from sickly, yellow plants and cross to emasculated flowers of a normal dark green-leaved plant. All progeny should have the normal dark green phenotype.

c.   The chloroplasts contain the green pigment chlorophyll and are the site of photosynthesis. A defect in the production of chlorophyll would give rise to all the stated defects.

22.   Maternal inheritance of chloroplasts results in the green-white color variegation observed in *Mirabilis*.
Cross 1:   variegated female × green male → variegated, green, or white progeny
Cross 2:   green female × variegated male → green progeny

In both crosses, the pollen (male contribution) contains no chloroplasts and thus does not contribute to the inheritance of this phenotype. Eggs from a variegated female plant can be of three types : contain only "green" chloroplasts, contain only "white" chloroplasts, or contain both (variegated). The offspring will have the phenotype associated with the egg's chloroplasts.

23.   The crosses are
Cross 1:   stop-start female × wild-type male → all stop-start progeny
Cross 2:   wild-type female × stop-start male → all wild-type progeny
                mtDNA is inherited only from the "female" in *Neurospora*.

24.   The genetic determinants of R and S are showing maternal inheritance and are therefore cytoplasmic. It is possible that the gene that confers resistance maps either to the mtDNA or cpDNA.

25.   The hypothesis is that the organism being tested is a dihybrid with independently assorting genes and that all progeny are of equal viability. The expected values would be that phenotypes occur with equal frequency. There are four phenotypes so there are 3 degrees of freedom.

$$\chi^2 = \sum (\text{observed-expected})^2/\text{expected}$$

$$\chi^2 = [(230{-}233)^2 + (210{-}233)^2 + (240{-}233)^2 + (250{-}233)^2]/233$$
= 3.75; *p* value between 0.1 and 0.5, nonsignificant; hypothesis cannot be rejected

26.   The hypothesis is that the organism being tested is a dihybrid with independently assorting genes and that all progeny are of equal viability. The expected values would be that phenotypes occur in a 9 : 3 : 3 : 1 ratio.. There are four phenotypes so there are 3 degrees of freedom.

$$\chi^2 = \sum (\text{observed-expected})^2/\text{expected}$$

$$\chi^2 = (178-180)^2/180 + (62-60)^2/60 + (56-60)^2/60 + (24-20)^2/20$$
$$= 1.156; p > 0.50, \text{ nonsignificant; hypothesis cannot be rejected}$$

27. When results of a cross are sex-specific, sex linkage should be considered. In moths, the heterogametic sex is actually the female while the male is the homogametic sex. Assuming that dark (D) is dominant to light (d), then the data can be explained by the dark male being heterozygous (D/d) and the dark female being hemizygous (D). All male progeny will inherit the D allele from their mother and therefore be dark while half the females will inherit D from their father (and be dark) and half will inherit d (and be light).

28. a. This inheritance pattern is diagnostic for organelle inheritance. These crosses indicate the mutant gene resides in the mitochondria. (Neurospora does not have chloroplasts.)

    b. All progeny from this cross will have the "maternal" trait stopper but the *nic3* allele should segregate 1:1 in the octad—four spores will be *stp* ; *nic3* and four spores will be *stp* ; *nic3*$^+$.

29. a. There should be nine classes—0, 1, 2, 3, 4, 5, 6, 7, 8 "doses"

    b. There should be 13 classes—0, 1, 2, 3, 4, 5, 6, 7, 8, 9, 10, 11, 12 "doses"

30. There are three ways for a self cross of this genotype to give rise to just one dose — one dose of R_1 and none of R_2 or R_3; one dose of R_2 and none of R_1 or R_3; or one dose of R_3 and none of R_1 or R_2. The chance of inheriting one dose of R_1 (from $R_1/r_1 \times R_1/r_1$) is $1/2$. The chance of no doses of R_2 (from $R_2/r_2 \times R_2/r_2$) is $1/4$ as is the chance of no doses of R_3 (from $R_3/r_3 \times R_3/r_3$). Therefore, the desired outcome of just one dose is $3 \times 1/2 \times 1/4 \times 1/4 = 3/32$.

31. a. Both the gametophyte and the sporophyte are closer in shape to the mother than the father. Note that a size increase occurs in each type of cross.

    b. Gametophyte and sporophyte morphology are affected by extranuclear factors. Leaf size may be a function of the interplay between nuclear genome contributions.

    c. If extranuclear factors are affecting morphology while nuclear factors are affecting leaf size, then repeated backcrosses could be conducted, using the hybrid as the female. This would result in the cytoplasmic information remaining constant while the nuclear information becomes increasingly like that of the backcross parent. Leaf morphology should therefore remain constant while leaf size would decrease toward the size of the backcross parent.

32. The goal here is to generate a plant with the cytoplasm of plant A and the nuclear genome predominantly of plant B. Remember that the cytoplasm is contributed by the egg only. So using plant A as the maternal parent, cross to B (as the paternal parent) and then backcross the progeny of this cross using plant B again as the paternal parent. Repeat for several generations until virtually the entire nuclear genome is from the B parent.

33. If the variegation is due to a chloroplast mutation, then the phenotype of the offspring will be controlled solely by the phenotype of the maternal parent. Look for flowers on white branches and test to see if they produce seeds that grow into all white plants regardless of the source of the pollen. To test whether a dominant nuclear mutation is responsible for the variegation, cross pollen from flowers on white branches to green plants (or flowers on green branches of same plant) and see if half the progeny are white and half are green (assuming the dominant mutation is heterozygous) or all white, if the dominant mutation is homozygous.

34. Progeny plants inherited only normal cpDNA (lane 1); only mutant cpDNA (lane 2); or both (lane 3). In order to get homoplasmic cpDNA (all chloroplasts containing the same DNA), seen in lanes 1 and 2, segregation of chloroplasts had to occur.

CHALLENGING PROBLEMS

35. **a.** Before beginning the specific problems, calculate the probabilities associated with each jar.

> jar 1 $p(R)$ 600/(600 + 400) = 0.6
> $p(W) = 400/(600 + 400) = 0.4$
>
> jar 2 $p(B) = 900/(900 + 100) = 0.9$
> $p(W)$ 100/(900 + 100) = 0.1
>
> jar 3 $p(G)$ 10/(10 + 990) = 0.01
> $p(W)$ 990/(10 + 990) = 0.99

(1) $p(R, B, G) = (0.6)(0.9)(0.01) = 0.0054$

(2) $p(W, W, W) = (0.4)(0.1)(0.99) = 0.0396$

(3) Before plugging into the formula, you should realize that, while white can come from any jar, red and green must come from specific jars (jar 1 and jar 3). Therefore, white must come from jar 2: $p(R, W, G) = (0.6)(0.1)(0.01) = 0.0006$

(4) $p(R, W, W) = (0.6)(0.1)(0.99) = 0.0594$

(5) There are three ways to satisfy this:

   R, W, W or W, B, W or W, W, G

  $= (0.6)(0.1)(0.99) + (0.4)(0.9)(0.99) + (0.4)(0.1)(0.01)$

   $= 0.0594 + 0.3564 + 0.0004 = 0.4162$

(6) At least one white is the same as 1 minus no whites:

  $p(\text{at least 1 W}) = 1 - p(\text{no W}) = 1 - p(R, B, G)$

   $= 1 - (0.6)(0.9)(0.01) = 1 - 0.0054 = 0.9946$

b. The cross is $R/r \times R/r$. The probability of red $(R/-)$ is $^3/_4$, and the probability of white (r/r) is $^1/_4$. Because only one white is needed, the only unacceptable result is all red.

In n trials, the probability of all red is $(^3/_4)^n$. Because the probability of failure must be no greater than 5 percent:

  $(^3/_4)^n < 0.05$

  $n > 10.41$, or 11 seeds

c. The $p(\text{failure}) = 0.8$ for each egg. Since all eggs are implanted simultaneously, the $p(\text{5 failures}) = (0.8)^5$. The $p(\text{at least one success}) = 1 - (0.8)^5 = 1 - 0.328 = 0.672$

36. Use the following symbols:

Gene function	Dominant allele	Recessive allele
color	R = red	r = yellow
loculed	L = two	l = many
height	H = tall	h = dwarf

The starting plants are pure-breeding, so their genotypes are

  red, two-loculed, dwarf R/R ; L/L ; h/h

and

  yellow, many-loculed, tall r/r ; l/l ; H/H.

The farmer wants to produce a pure-breeding line that is yellow, two-loculed, and tall, which would have the genotype r/r ; L/L ; H/H.

The two pure-breeding starting lines will produce an F_1 that will be R/r ; L/l ; H/h. By doing an $F_1 \times F_1$ cross, $^1/_{64}$ of the F_2 progeny should have the correct genotype ($^1/_4$ $r/r \times$ $^1/_4$ $L/L \times$ $^1/_4$ H/H). The probability of NOT getting that is $(1 - ^1/_{64})^n$, where n is number of progeny scored. For that to be less than 5 percent, n

= 191. So we need at least 191 progeny to start with, and by selecting yellow, two-loculed, and tall plants from these progeny, the known genotype will be r/r ; $L/-$; $H/-$. To identify how many of these are required for further testing (by test cross): the probability of being homozygous dominant for both (given that we are selecting only from those plants with dominant phenotypes) is $1/3 \times 1/3 = 1/9$. Therefore, the probability of a plant not being homozygous for both is $8/9$. We want the probability of all plants tested not being homozygous for both to be less than 5 percent, or $(8/9)^n < 0.05$. If $n = 26$, $p = .047$. So at least 26 of the yellow, two-loculed, and tall progeny should be test-crossed to a l/l; h/h parent to determine which are homozygous for the two dominant traits. (*Note:* several l/l ; h/h testers are likely to be recovered among the 191 F_2 progeny generated. So you will only need one test cross for each candidate.)

For each testcross, the plant will obviously be discarded if the testcross reveals a heterozygous state for the gene in question. If no recessive allele is detected, then the minimum number of progeny that must be examined to be 95 percent confident that the plant is homozygous, is based on the frequency of the dominant phenotype if heterozygous, which is $1/2$. In n progeny, the probability of obtaining all dominant progeny in a testcross, given that the plant is heterozygous, is $(1/2)^n$. To be 95 percent confident of homozygosity, the following formula is used, where 5 percent is the probability that it is not homozygous:

$$(1/2)^n = 0.05$$

$n = 4.3$, or 5 phenotypically dominant progeny must be obtained from each testcross to be 95 percent confident that the plant is homozygous.

37. **a.** Because each gene assorts independently, each probability should be considered separately and then multiplied together for the answer.

For (1) $1/2$ will be A, $3/4$ will be B, $1/2$ will be C, $3/4$ will be D, and $1/2$ will be E.

$$1/2 \times 3/4 \times 1/2 \times 3/4 \times 1/2 = 9/128$$

For (2) $1/2$ will be a, $3/4$ will be B, $1/2$ will be c, $3/4$ will be D, and $1/2$ will be e.

$$1/2 \times 3/4 \times 1/2 \times 3/4 \times 1/2 = 9/128$$

For (3) it is the sum of (1) and (2) = $9/128 + 9/128 = 9/64$

For (4) it is $1 - $(part 3) $= 1 - 9/64 = 55/64$

b. For (1) $1/2$ will be A/a, $1/2$ will be B/b, $1/2$ will be C/c, $1/2$ will be D/d, and $1/2$ will be E/e.

$$\frac{1}{2} \times \frac{1}{2} \times \frac{1}{2} \times \frac{1}{2} \times \frac{1}{2} = \frac{1}{32}$$

For (2) $\frac{1}{2}$ will be a/a, $\frac{1}{2}$ will be B/b, $\frac{1}{2}$ will be c/c, $\frac{1}{2}$ will be D/d, and $\frac{1}{2}$ will be e/e.

$$\frac{1}{2} \times \frac{1}{2} \times \frac{1}{2} \times \frac{1}{2} \times \frac{1}{2} = \frac{1}{32}$$

For (3) it is the sum of (1) and (2) = $\frac{1}{16}$.

For (4) it is $1 - (\text{part } 3) = 1 - \frac{1}{16} = \frac{15}{16}$

38. a. Cataracts appear to be caused by a dominant allele because affected people have affected parents. Dwarfism appears to be caused by a recessive allele because affected people have unaffected parents. Both traits appear to be autosomal.

b. Using A for cataracts, a for no cataracts, B for normal height, and b for dwarfism. The genotypes are:

III : a/a ; B/b, a/a ; B/b, A/a ; $B/-$, a/a ; $B/-$, A/a ; B/b, A/a ; B/b, a/a ; $B/-$, a/a ; $B/-$, a/a ; b/b

c. The mating is a/a ; b/b (IV-1) $\times$ $A/-$; $B/-$ (IV-5). Recall that the probability of a child's being affected by any disease is a function of the probability of each parent carrying the allele in question and the probability that one parent (for a dominant disorder) or both parents (for a recessive disorder) donate it to the child. Individual IV-1 is homozygous for these two genes, therefore, the only task is to determine the probabilities associated with individual IV-5.

The probability that individual IV-5 is heterozygous for dwarfism is $\frac{2}{3}$. Thus the probability that she has the b allele and will pass it to her child is $\frac{2}{3} \times \frac{1}{2} = \frac{1}{3}$.

The probability that individual IV-5 is homozygous for cataracts is $\frac{1}{3}$; the probability that she is heterozygous is $\frac{2}{3}$. If she is homozygous for the allele that causes cataracts, she must pass it to her child or if she is heterozygous for cataracts, she has a probability of $\frac{1}{2}$ of passing it to her child.

The probability that the first child is a dwarf with cataracts is the probability that the child inherits the A and b alleles from its mother which is $(\frac{1}{3} \times 1)(\frac{2}{3} \times \frac{1}{2}) + (\frac{2}{3} \times \frac{1}{2})(\frac{2}{3} \times \frac{1}{2}) = \frac{2}{9}$. Alternatively, you can calculate the chance of inheriting the b allele ($\frac{2}{3} \times \frac{1}{2}$) and not inheriting the a allele ($1 - \frac{1}{3}$) or $\frac{1}{3} \times \frac{2}{3} = \frac{2}{9}$.

The probability of having a phenotypically normal child is the probability that the mother donates the *a* and *B* (or not *b*) alleles, which is $(^2/_3 \times ^1/_2)(1-^1/_3) = ^2/_9$.

39.   a. and b. Begin with any two of the three lines and cross them. If, for example, you began with *a/a* ; *B/B* ; *C/C* × *A/A* ; *b/b* ; *C/C*, the progeny would all be *A/a* ; *B/b* ; *C/C*. Crossing two of these would yield:

9	*A/–* ; *B/–* ; *C/C*
3	*a/a* ; *B/–* ; *C/C*
3	*A/–* ; *b/b* ; *C/C*
1	*a/a* ; *b/b* ; *C/C*

The *a/a* ; *b/b* ; *C/C* genotype has two of the genes in a homozygous recessive state and occurs in $^1/_{16}$ of the offspring. If that were crossed with *A/A* ; *B/B* ; *c/c*, the progeny would all be *A/a* ; *B/b* ; *C/c*. Crossing two of them (or "selfing") would lead to a 27:9:9:9:3:3:3:1 ratio, and the plant occurring in $^1/_{64}$ of the progeny would be the desired *a/a* ; *b/b* ; *c/c*.

There are several different routes to obtaining *a/a* ; *b/b* ; *c/c*, but the one outlined above requires only four crosses.

40.   First, draw the pedigree.

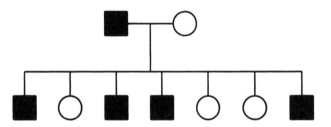

Let the genes be designated by the pigment produced by the normal allele : red pigment, *R;* green pigment, *G*; and blue pigment, *B*.

Recall that the sole X in males comes from the mother, while females obtain an X from each parent. Also recall that a difference in phenotype between sons and daughters is usually due to an X-linked gene. Because all the sons are colorblind and neither the mother nor the daughters are, the mother must carry a different allele for colorblindness on each X chromosome. In other words, she is heterozygous for both X-linked genes, and they are in repulsion: *R g/r G*. With regard to the autosomal gene, she must be *B/–*.

Because all the daughters are normal, the father, who is colorblind, must be able to complement the defects in the mother with regard to his X chromosome. Because he has only one X with which to do so, his genotype must be *R G/*Y ;

b/b. Likewise, the mother must be able to complement the father's defect, so she must be *B/B*.

The original cross is therefore

> P   *R g/r G ; B/B* × *R G/Y ; b/b*

> F$_1$ Females   Males
> *R g/R G ; B/b*   *R g/Y ; B/b*
> *r G/R G ; B/b*   *r G/Y ; B/b*

41. **a.**   The pedigree clearly shows maternal inheritance.

      b.   Most likely, the mutant DNA is mitochondrial.

42.   In the following schematic drawings, chromosomes (or chromatids) that are radioactive are indicated be the grains that would be observed after radioautography. After the second mitotic division, a number of outcomes are possible due to the random alignment and separation of the radioactive and non-radioactive chromatids.

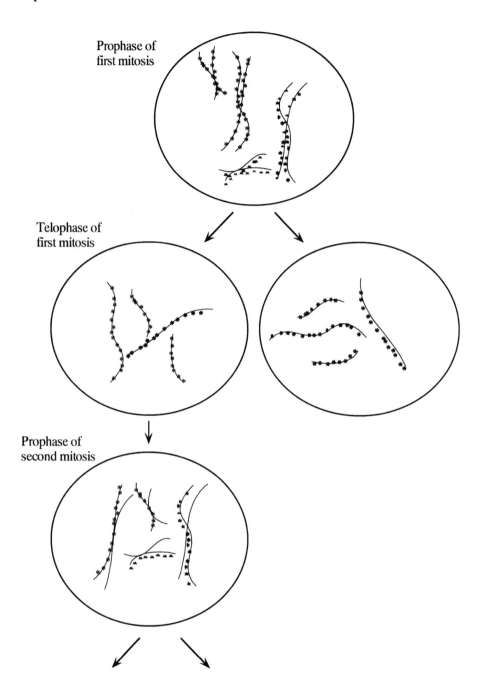

Prophase of
first mitosis

Telophase of
first mitosis

Prophase of
second mitosis

Telophase of
second mitosis

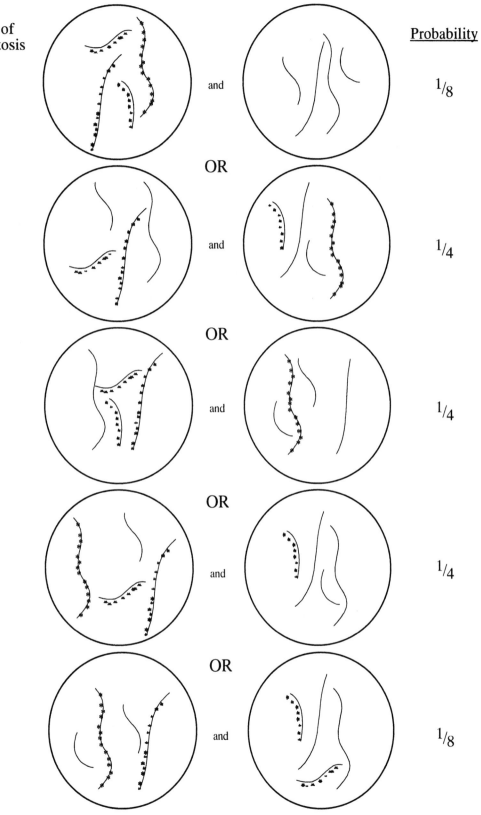

and   1/8

OR

and   1/4

OR

and   1/4

OR

and   1/4

OR

and   1/8

43.

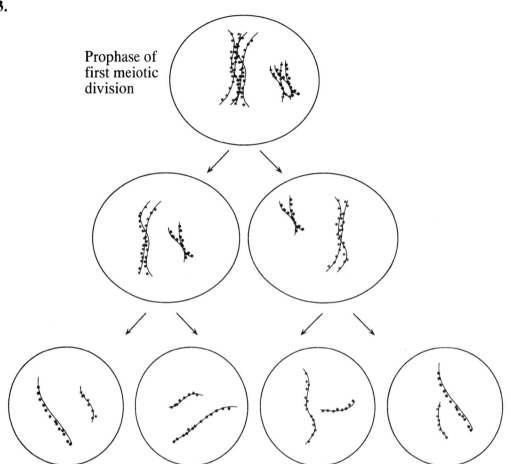

Prophase of
first meiotic
division

44. **a.** In a diploid cell, expect two chromosomes (a pair of homologs) to each have a single locus of radioactivity.

b. Expect many regions of radioactivity scattered throughout the chromosomes. The exact number and pattern would be dependent on the specific sequence in question, and where and how often it is present within the genome.

c. The multiple copies of the genes for ribosomal RNA are organized into large tandem arrays called nucleolar organizers (NO). Therefore, expect broader areas of radioactivity compared to (a). The number of these regions would equal the number of NO present in the organism.

d. Each chromosome end would be labeled by telomeric DNA.

e. The multiple repeats of this heterochromatic DNA are organized into large tandem arrays. Therefore, expect broader areas of radioactivity compared to (a). Also, there may be more than one area in the genome of the same simple repeat.

45. The following is meant to be examples of what is possible. It is also possible, for instance, that more than one band would be present in (a), depending on the position of the restriction sites within the sequence complementary to the probe used.

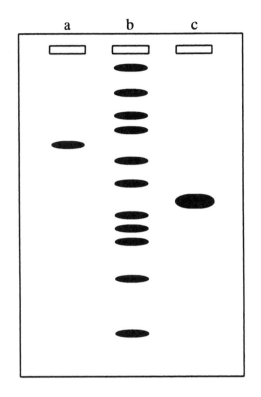

For (d) and (e), the specifics or where the DNA is cut relative to the telomeric DNA or heterochromatic DNA will effect what is observed. Assuming the restriction sites are not within the telomeric or heterochromatic DNA, then (d) will be similar to (b), and (c) will have one or several very large bands.

46. (1)   Impossible: the alleles of the same genes are on nonhomologous chromosomes
(2)   Meiosis II
(3)   Meiosis II
(4)   Meiosis II
(5)   Mitosis
(6)   Impossible: appears to be mitotic anaphase but alleles of sister chromatids are not identical
(7)   Impossible: too many chromosomes
(8)   Impossible: too many chromosomes
(9)   Impossible: too many chromosomes
(10) Meiosis I
(11) Impossible: appears to be meiosis of homozygous *a/a* ; *B/B*
(12) Impossible: the alleles of the same genes are on nonhomologous chromosomes

47. Recall that each cell has many mitochondria, each with numerous genomes. Also recall that cytoplasmic segregation is routinely found in mitochondrial mixtures within the same cell.

The best explanation for this pedigree is that the mother in generation I experienced a mutation in a single cell that was a progenitor of her egg cells (primordial germ cell). By chance alone, the two males with the disorder in the second generation were from egg cells that had experienced a great deal of cytoplasmic segregation prior to fertilization, while the two females in that generation received a mixture.

The spontaneous abortions that occurred for the first woman in generation II were the result of extensive cytoplasmic segregation in her primordial germ cells: aberrant mitochondria were retained. The spontaneous abortions of the second woman in generation II also came from such cells. The normal children of this woman were the result of extensive segregation in the opposite direction: normal mitochondria were retained. The affected children of this woman were from egg cells that had undergone less cytoplasmic segregation by the time of fertilization, so that they developed to term but still suffered from the disease.

48. **a.** Let B = brachydactylous, b = normal, T = taster and t = nontaster. The genotypes of the couple are B/b ; T/t for the male and b/b ; T/t for the female.

b. For all four children to be brachydactylous, the chance is $(1/2)^4 = 1/16$.

c. For none of the four children to be brachydactylous, the chance is $(1/2)^4 = 1/16$.

d. For all to be tasters, the chance is $(3/4)^4 = 81/256$.

e. For all to be nontasters, the chance is $(1/4)^4 = 1/256$.

f. For all to be brachydactylous tasters, the chance is $(1/2 \times 3/4)^4 = 81/4096$.

g. Not being a brachydactylous taster is the same 1– (the chance of being a brachydactylous taster) or $1 - (1/2 \times 3/4) = 5/8$. The chance that all four children are not brachydactylous tasters is $(5/8)^4 = 625/4096$.

h. The chance that at least one is a brachydactylous taster is $1 -$ (the chance of none being a brachydactylous taster) or $1 - (5/8)^4$.

49. For the following, S will signify cytoplasm of a male-sterile line and N will signify cytoplasm of a non-male-sterile line. *Rf* will signify the dominant nuclear restorer allele and *rf*, the recessive non-restorer allele.

a. If S *rf/rf* (male-sterile plants) are crossed with pollen from N *Rf/Rf* plants, the offspring will all be S *Rf/rf* and male fertile. If these offspring are then crossed with pollen from N *rf/rf* plants, half the offspring will be S *Rf/rf* (male-fertile) and half will be S *rf/rf* (male-sterile). The S cytoplasm will not be altered or affected even though the maternal offspring parent plant was *Rf/rf*.

b. The cross is S *rf/rf* × N *Rf/Rf* so all the progeny will be S *Rf/rf* and male-fertile.

c. The cross is S *Rf/rf* × N *rf/rf* so half the progeny will be S *Rf/rf* (male-fertile) and half will be S *rf/rf* (male-sterile).

d.   i.   The cross is S *rf-1/rf-1* ; *rf-2/rf-2* × N *Rf-1/rf-1* ; *Rf-2/rf-2*
The progeny will be: $^1/_4$ S *Rf-1/rf-1* ; *Rf-2/rf-2* (male-fertile)
$^1/_4$ S *Rf-1/rf-1* ; *rf-2/rf-2* (male-fertile)
$^1/_4$ S *rf-1/rf-1* ; *Rf-2/rf-2* (male-fertile)
$^1/_4$ S *rf-1/rf-1* ; *rf-2/rf-2* (male-sterile)

  ii.   The cross is S *rf-1/rf-1* ; *rf-2/rf-2* × N *Rf-1/Rf-1* ; *rf-2/rf-2*
The progeny will all be: S *Rf-1/rf-1* ; *rf-2/rf-2* (male-fertile)

  iii.   The cross is S *rf-1/rf-1* ; *rf-2/rf-2* × N *Rf-1/rf-1* ; *rf-2/rf-2*
The progeny will be: $^1/_2$ S *Rf-1/rf-1* ; *rf-2/rf-2* (male-fertile)
$^1/_2$ S *rf-1/rf-1* ; *rf-2/rf-2* (male-sterile)

  vi.   The cross is S *rf-1/rf-1* ; *rf-2/rf-2* × N *Rf-1/rf-1* ; *Rf-2/Rf-2*
The progeny will be: $^1/_2$ S *Rf-1/rf-1* ; *Rf-2/rf-2* (male-fertile)
$^1/_2$ S *rf-1/rf-1* ; *Rf-2/rf-2* (male-fertile)

4

Mapping Eukaryotic Chromosomes by Recombination

BASIC PROBLEMS

1. You perform the following cross and are told that the two genes are 10 m.u. apart.

$$A \; B/a \; b \; \times \; a \; b/a \; b$$

Among their progeny, 10 percent should be recombinant (*A b/a b* and *a B/a b*) and 90 percent should be parental (*A B/a b* and *a b/a b*). Therefore, *A B/a b* should represent $1/2$ of the parentals or 45 percent.

2. P   *A d / A d*   ×   *a D / a D*
F$_1$ *A d / a D*
F$_2$ 1 *A d / A d*   phenotype: A d
                2 *A d / a D*   phenotype: A D
                1 *a D / a D*   phenotype: a D

3. P   *R S/r s × R S/r s*

gametes $1/2 \, (1 - 0.35)$ *R S*
                    $1/2 \, (1 - 0.35)$ *r s*
                    $1/2 \, (0.35)$ *R s*
                    $1/2 \, (0.35)$ *r S*

F$_1$ genotypes   0.1056 *R S/R S*   0.1138 *r s/r S*
                    0.1056 *r s/r s*   0.1138 *r s/R s*
                    0.2113 *R S/r s*   0.0306 *R s/R s*
                    0.1138 *R S/r S*   0.0306 *r S/r S*
                    0.1138 *R S/R s*   0.0613 *R s/r S*

F$_1$ phenotypes 0.6058 R S
0.1056 r s
0.1444 R s
0.1444 r S

4.   The cross is *E/e · F/f × e/e · f/f*. If independent assortment exists, the progeny should be in a 1:1:1:1 ratio, which is not observed. Therefore, there is linkage. *E f* and *e F* are recombinants equaling one-third of the progeny. The two genes are 33.3 map units (m.u.) apart.

$$RF = 100\% \times 1/3 = 33.3\%$$

5.   Because only parental types are recovered, the two genes must be tightly linked and recombination must be very rare. Knowing how many progeny were looked at would give an indication of how close the genes are.

6.   The problem states that a female that is *A/a · B/b* is test crossed. If the genes are unlinked, they should assort independently and the four progeny classes should be present in roughly equal proportions. This is clearly not the case. The *A/a · B/b* and *a/a · b/b* classes (the parentals) are much more common than the *A/a · b/b* and *a/a · B/b* classes (the recombinants). The two genes are on the same chromosome and are 10 map units apart.

$$RF = 100\% \times (46 + 54)/1000 = 10\%$$

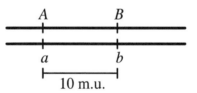

7.   The cross is *A/A · B/B × a/a · b/b*. The F$_1$ would be *A/a · B/b*.

   a.   If the genes are unlinked, all four progeny classes from the test cross (including *a/a · b/b*) would equal 25 percent.

   b.   With completely linked genes, the F$_1$ would produce only *A B* and *a b* gametes. Thus, there would be a 50 percent chance of having *a b/a b* progeny from a test cross of this F$_1$.

   c.   If the two genes are linked and 10 map units apart, 10 percent of the test cross progeny should be recombinants (and 90 percent should be parentals). Since the F$_1$ is *A B/a b*, *a b* is one of the parental classes (*A B* being the other) and should equal $^1/_2$ of the total parentals or 45 percent.

   d.   38 percent (see part c).

8. Meiosis is occurring in an organism that is *C d/c D*, ultimately producing haploid spores. The parental genotypes are *C d* and *c D*, in equal frequency. The recombinant types are *C D* and *c d*, in equal frequency. Eight map units means 8 percent recombinants. Thus, *C D* and *c d* will each be present at a frequency of 4 percent, and *C d* and *c D* will each be present at a frequency of $(100\% - 8\%)/2 = 46\%$.

   a.   4 percent

   b.   4 percent

   c.   46 percent

   d.   8 percent

9. To answer this question, you must realize that

   (1) One chiasma involves two of the four chromatids of the homologous pair so if 16 percent of the meioses have one chiasma, it will lead to 8 percent recombinants observed in the progeny (one half of the chromosomes of such a meiosis are still parental), and
   (2) Half of the recombinants will be *B r*, so the correct answer is 4 percent.

10. *Unpacking the Problem*

   1.   There is no correct drawing; any will do. Pollen from the tassels is placed on the silks of the females. The seeds are the F_1 corn kernels.

   2.   The +'s all look the same because they signify wild type for each gene. The information is given in a specific order, which prevents confusion, at least initially. However, as you work the problem, which may require you to reorder the genes, errors can creep into your work if you do not make sure that you reorder the genes for each genotype in exactly the same way. You may find it easier to write the complete genotype, p^+ instead of +, to avoid confusion.

   3.   The phenotype is purple leaves and brown midriff to seeds. In other words, the two colors refer to different parts of the organism.

   4.   There is no significance in the original sequence of the data.

   5.   A tester is a homozygous recessive for all genes being studied. It is used so that the meiotic products in the organism being tested can be seen directly in the phenotype of the progeny.

6.   The progeny phenotypes allow you to infer the genotypes of the plants. For example, *gre* stands for "green," the phenotype of $p^+/-$; *sen* stands for "virus-sensitive," the phenotype of $v^+/-$; and *pla* stands for "plain seed," the phenotype of $b^+/-$. In this test cross, all progeny have at least one recessive allele so the "gre sen pla" progeny are actually $p^+/p \cdot v^+/v \cdot b^+/b$.

7.   *Gametes* refers to the gametes of the two pure-breeding parents. F_1 *gametes* refers to the gametes produced by the completely heterozygous F_1 progeny. They indicate whether crossing-over and independent assortment have occurred. In this case, because there is either independent assortment or crossing-over, or both, the data indicate that the three genes are not so tightly linked that zero recombination occurred.

8.   The main focus is meiosis occurring in the F_1 parent.

9.   The gametes from the tester are not shown because they contribute nothing to the phenotypic differences seen in the progeny.

10.   Eight phenotypic classes are expected for three autosomal genes, whether or not they are linked, when all three genes have simple dominant-recessive relationships among their alleles. The general formula for the number of expected phenotypes is 2^n, where n is the number of genes being studied.

11.   If the three genes were on separate chromosomes, the expectation is a 1:1:1:1:1:1:1:1 ratio.

12.   The four classes of data correspond to the parentals (largest), two groups of single crossovers (intermediate), and double crossovers (smallest).

13.   By comparing the parentals with the double crossovers, gene order can be determined. The gene in the middle flips with respect to the two flanking genes in the double-crossover progeny. In this case, one parental is $+++$ and one double crossover is $p++$. This indicates that the gene for leaf color (p) is in the middle.

14.   If only two of the three genes are linked, the data can still be grouped, but the grouping will differ from that mentioned in (12) above. In this situation, the unlinked gene will show independent assortment with the two linked genes. There will be one class composed of four phenotypes in approximately equal frequency, which combined will total more than half the progeny. A second class will be composed of four phenotypes in approximately equal frequency, and the combined total will be less than half the progeny. For example, if the cross were $a\,b/+\,+\,;\,c/+\,\times\,a\,b/a\,b\,;\,c/c$, then the parental class (more frequent class) would have four components: $a\,b\,c$, $a\,b\,+$, $+\,+\,c$, and $+\,+\,+$. The recombinant class would be $a\,+\,c$, $a\,+\,+$, $+\,b\,c$, and $+\,b\,+$.

15. *Point* refers to locus. The usage does not imply linkage, but rather a testing for possible linkage. A four-point test cross would look like the following: $a/+ \cdot b/+ \cdot c/+ \cdot d/+ \times a/a \cdot b/b \cdot c/c \cdot d/d$.

16. A *recombinant* refers to an individual who has alleles inherited from two different grandparents, both of whom were the parents of the individual's heterozygous parent. Another way to think about this term is that in the recombinant individual's heterozygous parent, recombination took place among the genes that were inherited from his or her parents. In this case, the recombination took place in the F_1 and the recombinants are among the F_2 progeny.

17. The "recombinant for" columns refer to specific gene pairs and progeny that exhibit recombination between those gene pairs.

18. There are three "recombinant for" columns because three genes can be grouped in three different gene pairs.

19. *R* refers to recombinant progeny, and they contain different configurations of alleles than were present in their heterozygous parent.

20. Column totals indicate the number of progeny that experience crossing-over between the specific gene pairs. They are used to calculate map units between the two genes.

21. The diagnostic test for linkage is a recombination frequency of significantly less than 50 percent.

22. A map unit represents 1 percent crossing-over and is the same as a centimorgan.

23. In the tester, recombination cannot be detected in the gamete contribution to the progeny because the tester is homozygous. The F_1 individuals have genotypes fixed by their parents' homozygous state and, again, recombination cannot be detected in their phenotypes. Recombination between the P configurations occurs when the F_1 forms gametes, and is detected in the phenotypes of the F_2 progeny.

24. Interference $I = 1 -$ coefficient of coincidence $= 1 -$ (observed double crossovers/expected double crossovers). The expected double crossovers are equal to the (frequency of crossing-over in the first region, in this case between v and p) $\times$ (frequency of crossing-over in the second region, between p and b) $\times$ number of progeny.

25. If the three genes are not all linked, then interference cannot be calculated.

26. A great deal of work is required to obtain 10,000 progeny in corn because each seed on a cob represents one progeny. Each cob may contain as many as 200 seeds. While seed characteristics can be assessed at the cob stage, for other characteristics such as leaf color, viral sensitivity, and midriff color, each seed must be planted separately and assessed after germination and growth. The bookkeeping task is also enormous.

Solution to the Problem

a. The three genes are linked.

b. Comparing the parentals (most frequent) with the double crossovers (least frequent), the gene order is $v\ p\ b$. There were 2200 recombinants between v and p, and 1500 between p and b. The general formula for map units is

m.u. = 100%(number of recombinants)/total number of progeny
Therefore, the map units between v and p = 100%(2200)/10,000 = 22 m.u., and the map units between p and b = 100%(1500)/10,000 = 15 m.u.
The map is

$$v \underset{\text{22 m.u.}}{\rule{4cm}{0.4pt}} p \underset{\text{15 m.u.}}{\rule{3cm}{0.4pt}} b$$

c. I = 1 – observed double crossovers/expected double crossovers
= 1 – 132/(0.22)(0.15)(10,000)
= 1 – 0.4 = 0.6

11.   P $a\ b\ c/a\ b\ c \times a^+\ b^+\ c^+/a^+\ b^+\ c^+$

F$_1$ $a^+\ b^+\ c^+/a\ b\ c \times a^+\ b^+\ c^+/a\ b\ c$

F$_2$		
	1364	$a^+\ b^+\ c^+$
	365	$a\ b\ c$
	87	$a\ b\ c^+$
	84	$a^+\ b^+\ c$
	47	$a\ b^+\ c^+$
	44	$a^+\ b\ c$
	5	$a\ b^+\ c$
	4	$a^+\ b\ c^+$

This problem is somewhat simplified by the fact that recombination does not occur in male *Drosophila*. Also, only progeny that received the $a\ b\ c$ chromosome from the male will be distinguishable among the F$_2$ progeny.

a. Because you cannot distinguish between $a^+ b^+ c^+/a^+ b^+ c^+$ and $a^+ b^+ c^+/a\ b\ c$, use the frequency of $a\ b\ c/a\ b\ c$ to estimate the frequency of $a^+ b^+ c^+$ (parental) gametes from the female.

parentals	730	(2×365)
CO a–b:	91	$(a\ b^+ c^+, a^+ b\ c = 47 + 44)$
CO b–c:	171	$(a\ b\ c^+, a^+ b^+ c = 87 + 84)$
DCO:	9	$(a^+ b\ c^+, a\ b^+ c = 4 + 5)$
	1001	

a–b: $100\%(91 + 9)/1001 = 10$ m.u.
b–c: $100\%(171 + 9)/1001 = 18$ m.u.

b. Coefficient of coincidence = (observed DCO)/(expected DCO)
    $= 9/[(0.1)(0.18)(1001)] = 0.5$

12. a. By comparing the two most frequent classes (parentals: $an\ br^+ f^+$, $an^+ br$ f) to the least frequent classes (DCO: $an^+ br\ f^+$, $an\ br^+ f$), the gene order can be determined. The gene in the middle switches with respect to the other two (the order is $an\ f\ br$). Now the crosses can be written fully.

P $an\ f^+ br^+/an\ f^+ br^+ \times an^+ f\ br/an^+ f\ br$

F$_1$ $an^+ f\ br/an\ f^+ br^+ \times an\ f\ br/an\ f\ br$

F$_2$	355	$an\ f^+ br^+/an\ f\ br$	parental
	339	$an^+ f\ br/an\ f\ br$	parental
	88	$an^+ f^+ br^+/an\ f\ br$	CO an–f
	55	$an\ f\ br/an\ f\ br$	CO an–f
	21	$an^+ f\ br^+/an\ f\ br$	CO f–br
	17	$an\ f^+ br/an\ f\ br$	CO f–br
	2	$an^+ f^+ br/an\ f\ br$	DCO
	2	$an\ f\ br^+/an\ f\ br$	DCO
	879		

b. an–f: $100\% (88 + 55 + 2 + 2)/879 = 16.72$ m.u.
   f–br: $100\% (21 + 17 + 2 + 2)/879 = 4.78$ m.u.

c. Interference = 1 – (observed DCO/expected DCO)
    $= 1 - 4/(0.1672)(0.0478)(879) = 0.431$

13. By comparing the most frequent classes (parental: $+ v \, lg$, $b + +$) with the least frequent classes (DCO: $+ + +$, $b \, v \, lg$) the gene order can be determined. The gene in the middle switches with respect to the other two, yielding the following sequence: $v \, b \, lg$. Now the cross can be written

P $v \, b^+ \, lg / v^+ \, b \, lg^+ \times v \, b \, lg / v \, b \, lg$

F_1	305	$v \, b^+ \, lg / v \, b \, lg$	parental
	275	$v^+ \, b \, lg^+ / v \, b \, lg$	parental
	128	$v^+ \, b \, lg / v \, b \, lg$	CO b–lg
	112	$v \, b^+ \, lg^+ / v \, b \, lg$	CO b–lg
	74	$v^+ \, b^+ \, lg / v \, b \, lg$	CO v–b
	66	$v \, b \, lg^+ / v \, b \, lg$	CO v–b
	22	$v^+ \, b^+ \, lg^+ / v \, b \, lg$	DCO
	18	$v \, b \, lg / v \, b \, lg$	DCO

v–b: $100\%(74 + 66 + 22 + 18)/1000 = 18.0$ m.u.
b–lg: $100\%(128 + 112 + 22 + 18)/1000 = 28.0$ m.u.

c.c.   = observed DCO/expected DCO = $(22 + 18)/(0.28)(0.18)(1000) = 0.79$

14. Let F = fat, L = long tail, and Fl = flagella. The gene sequence is $F \, L \, Fl$ (compare most frequent with least frequent). The cross is

P $F \, L \, Fl / f \, l \, fl \times f \, l \, fl / f \, l \, fl$

F_1	398	$F \, L \, Fl / f \, l \, fl$	parental
	370	$f \, l \, fl / f \, l \, fl$	parental
	72	$F \, L \, fl / f \, l \, fl$	CO L–Fl
	67	$f \, l \, Fl / f \, l \, fl$	CO L–Fl
	44	$f \, L \, FL / f \, l \, fl$	CO F–L
	35	$F \, l \, fl / f \, l \, fl$	CO F–L
	9	$f \, L \, fl / f \, l \, fl$	DCO
	5	$F \, l \, Fl / f \, l \, fl$	DCO

L–Fl: $100\%(72 + 67 + 9 + 5)/1000 = 15.3$ m.u.
F–L: $100\%(44 + 35 + 9 + 5)/1000 = 9.3$ m.u.

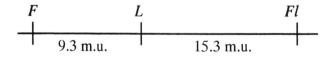

15. a.   The hypothesis is that the genes are not linked. Therefore, a 1:1:1:1 ratio is expected.

b. $\chi^2 = (54–50)^2/50 + (47–50)^2/50 + (52–50)^2/50 + (47–50)^2/50$
$= 0.32 + 0.18 + 0.08 + 0.18 = 0.76$

c. With 3 degrees of freedom, the p value is between 0.50 and 0.90.

d. Between 50 percent and 90 percent of the time values this extreme from the prediction would be obtained by chance alone.

e. Accept the initial hypothesis.

f. Because the χ^2 value was insignificant, we conclude the two genes are assorting independently. The genotypes of all individuals are

P $dp^+/dp^+ ; e/e \times dp/dp ; e^+/e^+$

F_1 $dp^+/dp ; e^+/e$

tester $dp/dp ; e/e$

progeny   long, ebony $dp^+/dp ; e/e$
             long, gray $dp^+/dp ; e^+/e$
             short, gray $dp/dp ; e^+/e$
             short, ebony $dp/dp ; e/e$

16. a.

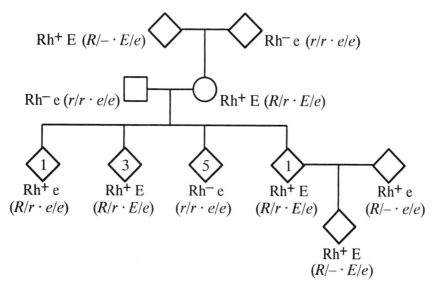

b. Yes

c. Dominant

d. As drawn, the pedigree hints at linkage. If unlinked, expect that the phenotypes of the 10 children should be in a 1:1:1:1 ratio of Rh$^+$ E, Rh$^+$ e,

Rh⁻ E, and Rh⁻ e. There are actually five Rh⁻ e, four Rh⁺ E, and one Rh⁺ e. If linked, this last phenotype would represent a recombinant, and the distance between the two genes would be 100%(1/10) = 10 m.u. However, there is just not enough data to strongly support that conclusion.

17. Assume there is no linkage. (This is your hypothesis. If it can be rejected, the genes are linked.) The expected values would be that genotypes occur with equal frequency. There are four genotypes in each case ($n = 4$) so there are 3 degrees of freedom.

$$\chi^2 = \sum (\text{observed} - \text{expected})^2/\text{expected}$$

Cross 1: $\chi^2 = [(310{-}300)^2 + (315{-}300)^2 + (287{-}300)^2 + (288{-}300)^2]/300$
$= 2.1266; p > 0.50$, nonsignificant; hypothesis cannot be rejected

Cross 2: $\chi^2 = [(36{-}30)^2 + (38{-}30)^2 + (23{-}30)^2 + (23{-}30)^2]/30$
$= 6.6; p > 0.10$, nonsignificant; hypothesis cannot be rejected

Cross 3: $\chi^2 = [(360{-}300)^2 + (380{-}300)^2 + (230{-}300)^2 + (230{-}300)^2]/300$
$= 66.0; p < 0.005$, significant; hypothesis must be rejected

Cross 4: $\chi^2 = [(74{-}60)^2 + (72{-}60)^2 + (50{-}60)^2 + (44{-}60)^2]/60$
$= 11.60; p < 0.01$, significant; hypothesis must be rejected

18. a. Both disorders must be recessive to yield the patterns of inheritance that are observed. Notice that only males are affected, strongly suggesting X linkage for both disorders. In the first pedigree there is a 100 percent correlation between the presence or absence of both disorders, indicating close linkage. In the second pedigree, the presence and absence of both disorders are inversely correlated, again indicating linkage. In the first pedigree, the two defective alleles must be cis within the heterozygous females to show 100 percent linkage in the affected males, while in the second pedigree the two defective alleles must be trans within the heterozygous females.

b. and c. Let *a* stand for the allele giving rise to steroid sulfatase deficiency (vertical bar) and *b* stand for the allele giving rise to ornithine transcarbamylase deficiency (horizontal bar). Crossing over cannot be detected without attaching genotypes to the pedigrees. When this is done, it can be seen that crossing-over need not occur in either of the pedigrees to give rise to the observations.

First pedigree:

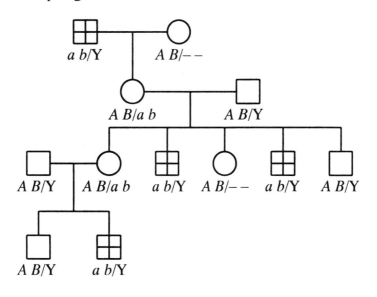

Second pedigree:

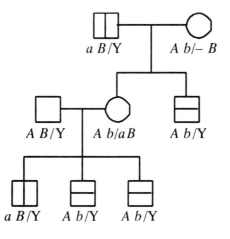

19. a. Note that only males are affected by both disorders. This suggests that both are X-linked recessive disorders. Using *p* for protan and *P* for non-protan, and *d* for deutan and *D* for non-deutan, the inferred genotypes are listed on the pedigree below. The Y chromosome is shown, but the X is represented by the alleles carried.

 b. Individual II-2 must have inherited both disorders in the trans configuration (on separate chromosomes). Therefore, individual III-2 inherited both traits as the result of recombination (crossing-over) between his mother's X chromosomes.

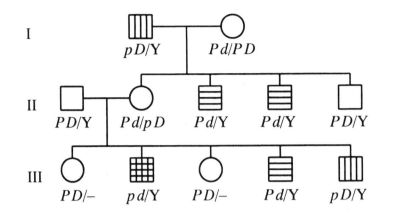

c. Because both genes are X-linked, this represents crossing-over. The progeny size is too small to give a reliable estimate of recombination.

20. **a. and b.** Again, the best way to determine whether there is linkage is through chi-square analysis, which indicates that it is highly unlikely that the three genes assort independently. To determine linkage by simple inspection, look at gene pairs. Because this is a test cross, independent assortment predicts a 1:1:1:1 ratio.

Comparing shrunken and white, the frequencies are

+	+	(113 + 4)/total
s	wh	(116 + 2)/total
+	wh	(2708 + 626)/total
s	+	(2538 + 601)/total

There is not independent assortment between shrunken and white, which means that there is linkage.

Comparing shrunken and waxy, the frequencies are

+	+	(626 + 4)/total
s	wa	(601 + 2)/total
+	wa	(2708 + 113)/total
s	+	(2538 + 116)total

There is not independent assortment between shrunken and waxy, which means that there is linkage.

Comparing white and waxy, the frequencies are

+	+	(2538 + 4)/total
wh	wa	(2708 + 2)/total
wh	+	(626 + 116)/total
+	wa	(601 + 113)/total

There is not independent assortment between waxy and white, which means that there is linkage.

Because all three genes are linked, the strains must be $+ s +/wh + wa$ and $wh\ s\ wa/wh\ s\ wa$ (compare most frequent, parentals, to least frequent, double crossovers, to obtain the gene order). The cross can be written as

P $+ s +/wh + wa \times wh\ s\ wa/wh\ s\ wa$

F_1 as in problem

Crossovers between white and shrunken and shrunken and waxy are

113	601
116	626
4	4
2	2
235	1233

Dividing by the total number of progeny and multiplying by 100 percent yields the following map:

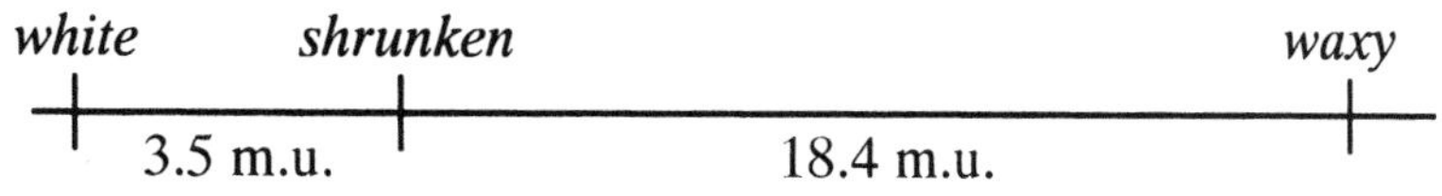

white	*shrunken*		*waxy*
+	+		+
	3.5 m.u.	18.4 m.u.	

c.   Interference   = 1 − (observed double crossovers/expected double crossovers)
          = 1 − 6/(0.035)(0.184)(6,708) = 0.86

21. **a.** The results of this cross indicate independent assortment of the two genes. This might be diagrammed as

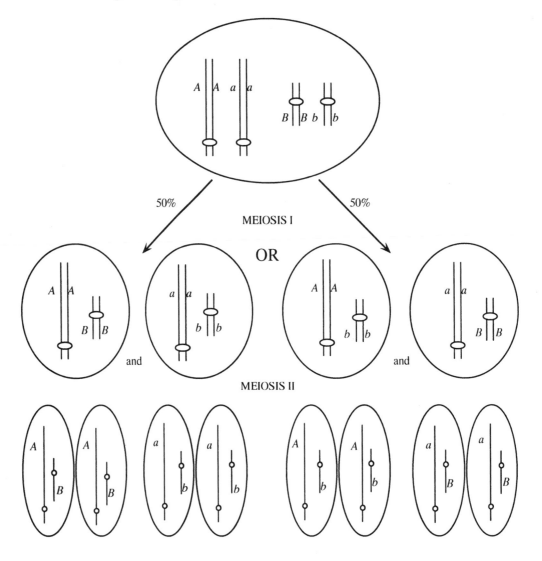

b. The results of this cross indicate that the two genes are linked and 10 m.u. apart. Further, the recessive alleles are in repulsion in the dihybrid (*C d/c D* × *c d/c d*). This might be diagrammed as

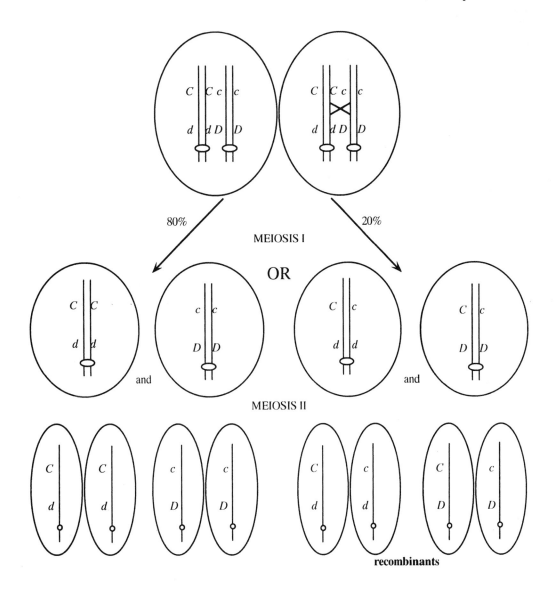

22. **a.**   If the genes are unlinked, the cross becomes

P hyg/hyg ; her/her × hyg^+/hyg^+ ; her^+/her^+

F_1 hyg^+/hyg ; her^+/her × hyg^+/hyg ; her^+/her

F_2 $9/16$ $hyg^+/-$; $her^+/-$
                        $3/16$ $hyg^+/-$; her/her
                        $3/16$ hyg/hyg ; $her^+/-$
                        $1/16$ hyg/hyg ; her/her

So only $1/16$ (or 6.25 percent) of the seeds would be expected to germinate.

b. and c. No. More than twice the expected seeds germinated so assume the genes are linked. The cross then becomes

P   *hyg her/hyg her* × *hyg⁺ her⁺/hyg⁺ her⁺*

F_1 *hyg⁺ her⁺/hyg her* × *hyg⁺ her⁺/hyg her*

F_2 13 percent   *hyg her/hyg her*

Because this class represents the combination of two parental chromosomes, it is equal to
$$p(hyg\ her) \times p(hyg\ her) = (^1/_2\ \text{parentals})^2 = 0.13$$
and
$$\text{parentals} = 0.72 \qquad \text{so recombinants} = 1 - 0.72 = 0.28$$

Therefore, a test cross of *hyg⁺ her⁺/hyg her* should give

36%	*hyg⁺ her⁺/hyg her*
36%	*hyg her/hyg her*
14%	*hyg⁺ her/hyg her*
14%	*hyg her⁺/hyg her*

and 36 percent of the progeny should grow (the *hyg her/hyg her* class).

23. Comparing independent assortment (RF = 0.5) to a recombination frequency of 34 percent (RF = 0.34)

RF	0.5	0.34
P	0.25	0.33
P	0.25	0.33
R	0.25	0.17
R	0.25	0.17

The probability of obtaining these results, assuming independent assortment, will be equal to

0.25 × 0.25 × 0.25 × 0.25 × 0.25 × 0.25 × B = 0.00024B (where B = number of possible birth orders for four parental and two recombinant progeny)

For an RF = 0.34, the probability of this result will be equal to

0.33 × 0.17 × 0.33 × 0.33 × 0.17 × 0.33 × B = 0.00034B

The ratio of the two = 0.00034B/0.00024B = 1.42 and the Lod = 0.15

24. **a.** Meiosis I (crossing-over has occurred between all genes and their centromeres)

 b. Impossible

 c. Meiosis I (crossing-over has occurred between gene B and its centromere)

 d. Meiosis I

 e. Meiosis II

 f. Meiosis II (crossing-over has occurred between genes A and B and their centromeres)

 g. Meiosis II

 h. Impossible

 i. Mitosis

 j. Impossible

25. **a.**   *al-2⁺*   *al-2*

 b. The 8 percent value can be used to calculate the distance between the gene and the centromere. That distance is ½ the percentage of second-division segregation, or 4 percent.

26. **a.** *arg-6 · al-2, arg-6 · al-2, arg-6⁺ · al-2⁺, arg-6⁺ · al-2⁺*

 b. *arg-6⁺ · al-2⁺, arg-6⁺ · al-2, arg-6 · al-2⁺, arg-6 · al-2*

 c. *arg-6⁺ · al-2, arg-6⁺ · al-2, arg-6 · al-2⁺, arg-6 · al-2⁺*

27. The formula for this problem is $f(i) = e^{-m}m^i/i!$ where $m = 2$ and $i = 0, 1,$ or 2.

 a. $f(0) = e^{-2}2^0/0! = e^{-2} = 0.135$ or 13.5%

 b. $f(1) = e^{-2}2^1/1! = e^{-2}(2) = 0.27$ or 27%

 c. $f(2) = e^{-2}2^2/2! = e^{-2}(2) = 0.27$ or 27%

28. ***Unpacking the Problem***

 1. Fungi are generally haploid.

2. There are four pairs, or eight ascospores, in each ascus. One member of each pair is presented in the data.

3. A mating type in fungi is analogous to gender in humans, in that the mating types of two organisms must differ in order to have a mating that produces progeny. Mating type is determined experimentally simply by seeing if progeny result from specific crosses.

4. The mating types *A* and *a* do not indicate dominance and recessiveness. They simply symbolize the mating-type difference.

5. *arg-1* indicates that the organism requires arginine for growth. Testing for the genotype involves isolating nutritional mutants and then seeing if arginine supplementation will allow for growth.

6. *arg-1*$^+$ indicates that the organism is wild type and does not require supplemental arginine for growth.

7. Wild type refers to the common form of an organism in its natural population.

8. Mutant means that, for the trait being studied, an organism differs from the wild type.

9. The actual function of the alleles in this problem does not matter in solving the problem.

10. Linear octad analysis refers to the fact that the ascospores in each ascus are in a linear arrangement that reflects the order in which the two meiotic divisions and a subsequent mitotic division occurred to produce them. By tracking traits and correlating them with position, it is possible to detect crossing-over events that occurred at the tetrad (four-strand, homologous pairing) stage prior to the two meiotic divisions. Since the mitotic division occurs last, and mitotic sister spores are typically identical, the mitotic sisters may be treated as a pair of identical twins and are listed only once in the diagram, so the octad is treated as a tetrad.

11. Linear tetrad analysis allows for the mapping of centromeres in relation to genes, which cannot be done with unordered tetrad analysis.

12. A cross is made in *Neurospora* by placing the two organisms in the same test tube or Petri dish and allowing them to grow. Gametes develop and fertilization, followed by meiosis, mitosis, and ascus formation, occurs. The asci are isolated, and the ascospores are dissected out of them with the aid of a microscope. The ascus has an octad, or eight spores, within it, and the spores are arranged in four (tetrad) pairs.

13. Meiosis occurs immediately following fertilization in *Neurospora*.

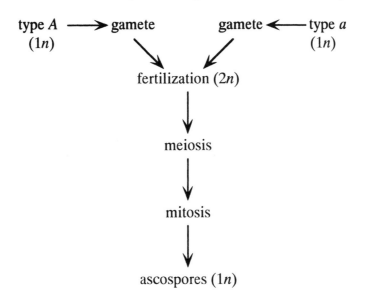

14. Meiosis produced the ascospores that were analyzed.

15. The cross is $A \cdot arg\text{-}1 \times a \cdot arg\text{-}1^+$.

16. Although there are eight ascospores, they occur in pairs. Each pair represents one chromatid of the originally paired chromosomes. By convention, both members of a pair are represented by a single genotype.

17. The seven classes represent the seven types of outcomes. The specific outcomes can be classified as follows:

Class	1	2	3	4	5	6	7
outcome	PD	NPD	T	T	PD	NPD	T
A/a	I	I	I	II	II	II	II
$arg\text{-}1^+/arg\text{-}1$	I	I	II	I	II	II	II

where PD = parental ditype, NPD = nonparental ditype, T = tetratype, I = first-division segregation, and II = second-division segregation.

Other classes can be detected, but they indicate the same underlying process. For example, the following three asci are equivalent.

1	2	3
$A\ arg$	$A\ arg^+$	$A\ arg^+$
$A\ arg^+$	$A\ arg$	$A\ arg$

a arg	*a arg*	*a arg*$^+$
a arg$^+$	*a arg*$^+$	*a arg*

In the first ascus, a crossover occurred between chromatids 2 and 3, while in the second ascus it occurred between chromatids 1 and 3, and in the third ascus the crossover was between chromatids 1 and 4. A fourth equivalent ascus would contain a crossover between chromatids 2 and 3. All four indicate a crossover between the second gene and its centromere and all are tetratypes.

18. This is exemplified in the answer to (17) above.

19. The class is identical with class 1 in the problem, but inverted. These are included in class 1 in the table.

20. Linkage arrangement refers to the relative positions of the two genes and the centromere along the length of the chromosome.

21. A genetic interval refers to the region between two loci, whose size is measured in map units.

22. The problem does not specify whether the two loci are on separate chromosomes or are on the same chromosome. The general formula for calculating the distance of a locus to its centromere is to measure the percentage of tetrads that show second-division segregation patterns for that locus and divide by two. You are supposed to determine whether the genes share a common centromere or are associated with different centromeres.

23. Recall that there are eight ascospores per ascus. By inspection, the frequency of recombinant *A arg-1*$^+$ ascospores is 4(125) + 2(100) + 2(36) + 4(4) + 2(6) = 800. There is also the reciprocal recombinant genotype *a arg-1*.

24. Class 1 is parental; class 2 is nonparental ditype. Because they occur at equal frequencies, the two genes are not linked.

Solution to the Problem

a. The cross is *A · arg-1* × *a · arg-1*$^+$. Use the classification of asci in part (17) above. First decide if the two genes are linked by using the formula PD>>NPD, when the genes are linked, while PD = NPD when they are not linked. PD = 127 + 2 = 129 and NPD = 125 + 4 = 129, which means that the two genes are not linked. Alternatively,

$$RF = 100\%(\tfrac{1}{2}T + NPD)/\text{total asci}$$

$$= 100\%[(^{1}/_{2})(100 + 36 + 6) + (125 + 4)]/400 = 50\%.$$

Next calculate the distance between each gene and its centromere using the formula $RF = 100\%(^{1}/_{2}$ number of tetrads exhibiting M_{II} segregation)/(total number of asci).

$$A\text{–centromere} = 100\%(^{1}/_{2})(36 + 2 + 4 + 6)/400$$

$$= 100\%(24/400) = 6 \text{ m.u.}$$

$$arg^{+}\text{–centromere} = 100\%(^{1}/_{2})(100 + 2 + 4 + 6)/400$$

$$= 100\%(56/400) = 14 \text{ m.u.}$$

b. Class 6 can be obtained if a single crossover occurred between chromatids 2 and 3 between each gene and its centromere.

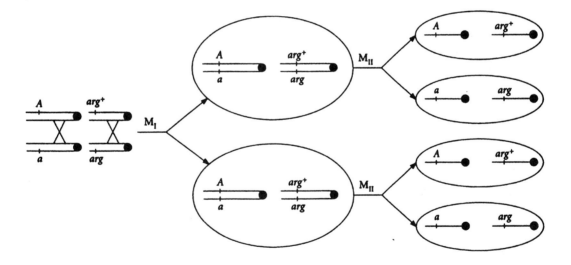

29. Before beginning this problem, classify all asci as PD, NPD, or T and determine whether there is M_I or M_{II} segregation for each gene:

	\multicolumn{7}{c}{Asci type}						
	1	2	3	4	5	6	7
type	PD	NPD	T	T	PD	NPD	T
gene a	I	I	I	II	II	II	II
gene b	I	I	II	I	II	II	II

If PD >> NPD, linkage is indicated. The distance between a gene and its centromere = 100% $(1/2)(M_{II})$/total. The distance between two genes = 100% $(1/2T + NPD)$/total.

Cross 1: PD = NPD and RF = 50%; the genes are not linked.
a–centromere: 100% $(1/2)(0)$/100 = 0 m.u. Gene *a* is close to the centromere.
b–centromere: 100% $(1/2)(32)$/100 = 16 m.u.

Cross 2: PD >> NPD; the genes are linked.
a–*b*: 100% $[1/2(15) + 1]$/100 = 8.5 m.u.
a–centromere: 100% $(1/2)(0)$/100 = 0 m.u. Gene *a* is close to the centromere.
b–centromere: 100% $(1/2)(15)$/100 = 7.5 m.u.

Cross 3: PD >> NPD; the genes are linked.
a–*b*: 100% $[1/2(40) + 3]$/100 = 23 m.u.
a–centromere: 100% $(1/2)(2)$/100 = 1 m.u.
b–centromere: 100% $(1/2)(40 + 2)$/100 = 21 m.u.

Cross 4: PD > > NPD; the genes are linked.
a–*b*: 100% $[1/2(20) + 1]$/100 = 11 m.u.
a–centromere: 100% $(1/2)(10)$/100 = 5 m.u.
b–centromere: 100% $(1/2)(18 + 8 + 1)$/100 = 13.5 m.u.

Cross 5: PD = NPD (and RF = 49%); the genes are not linked.
a–centromere: 100% $(1/2)(22 + 8 + 10 + 20)$/99 = 30.3 m.u.
b–centromere: 100% $(1/2)(24 + 8 + 10 + 20)$/99 = 31.3 m.u.
These values are approaching the 67 percent theoretical limit of loci exhibiting M_{II} patterns of segregation and should be considered cautiously.

Cross 6: PD >> NPD; the genes are linked.
a–*b*: 100% $[1/2(1 + 3 + 4) + 0]$/100 = 4 m.u.
a–centromere: 100% $(1/2)(3 + 61 + 4)$/100 = 34 m.u.
b–centromere: 100% $(1/2)(1 + 61 + 4)$/100 = 33 m.u.

These values are at the 67 percent theoretical limit of loci exhibiting M_{II} patterns of segregation and therefore both loci can be considered unlinked to the centromere.

Cross 7: PD >> NPD; the genes are linked.
a–*b*: 100% $[1/2(3 + 2) + 0]$/100 = 2.5 m.u.
a–centromere: 100% $(1/2)(2)$/100 = 1 m.u.
b–centromere: 100% $(1/2)(3)$/100 = 1.5 m.u.

Cross 8: PD = NPD; the genes are not linked.
a–centromere: 100% $(^1/_2)(22 + 12 + 11 + 22)/100 = 33.5$ m.u.
b–centromere: 100% $(^1/_2)(20 + 12 + 11 + 22)/100 = 32.5$ m.u.
Same as cross 5.

Cross 9: PD >> NPD; the genes are linked.
a–b: 100% $[^1/_2(10 + 18 + 2) + 1]/100 = 16$ m.u.
a–centromere: 100% $(^1/_2)(18 + 1 + 2)/100.= 10.5$ m.u.
b–centromere: 100% $(^1/_2)(10 + 1 + 2)/100 = 6.5$ m.u.

Cross 10: PD = NPD; the genes are not linked.
a–centromere: 100% $(^1/_2)(60 + 1 + 2 + 5)/100 = 34$ m.u.
b–centromere: 100% $(^1/_2)(2 + 1 + 2 + 5)/100 = 5$ m.u.

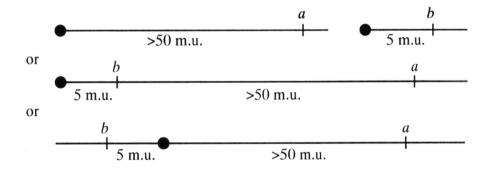

Cross 11: PD = NPD; the genes are not linked.
a–centromere: 100% $(^1/_2)(0)/100 = 0$ m.u.
b–centromere: 100% $(^1/_2)(0)/100 = 0$ m.u.

30. The number of recombinants is equal to NPD + $^1/_2$T. The uncorrected map distance is based on RF = (NPD + $^1/_2$T)/total. The corrected map distance, based on the Perkins' formula, is RF = 50(T + 6NPD)/total. The formula for the Haldane mapping function, is RF = ½$(1 - e^{-m})$. To convert to map distance, m, the measure of crossover frequency, is multiplied by 50 percent, to give the corrected map distance.

Cross 1:
recombinant frequency = 4% + $^1/_2$(45%) = 26.5%
uncorrected map distance = [4% + $^1/_2$(45%)]/100% = 26.5 m.u.
corrected map distance, Perkins' formula = 50[45% + 6(4%)]/100% = 34.5 m.u.
The formula for the Haldane mapping function is RF = ½$(1 - e^{-m})$. Solving for this situation, $e^{-m} = 1 - (2 \times 0.265) = 0.47$ so $m = 0.755$. To convert to map distance, m, the measure of crossover frequency, is multiplied by 50 percent, to give a corrected map distance of 37.8%.

Cross 2:

recombinant frequency $= 2\% + \frac{1}{2}(34\%) = 19\%$

uncorrected map distance $= [2\% + \frac{1}{2}(34\%)]/100\% = 19$ m.u.

corrected map distance, Perkins' formula $= 50[34\% + 6(2\%)]/100\% = 23$ m.u.

corrected map distance, using mapping function $= 23.9$ m.u.

Cross 3:

recombinant frequency $= 5\% + \frac{1}{2}(50\%) = 30\%$

uncorrected map distance $= [5\% + \frac{1}{2}(50\%)]/100\% = 30$ m.u.

corrected map distance, Perkins' formula $= 50[50\% + 6(5\%)]/100\% = 40$ m.u.

corrected map distance, using mapping function $= 45.8$ m.u.

31. The *leu3* gene is centromere-linked and always segregates at the first division (M_I patterns). The *cys2* gene is at a distance from its centromere and recombination between these respective locations will result in a second-division segregation (M_{II}) pattern. For this, there are four patterns of spores, all equally likely.

cys		cys		+		+
+	and	+	and	cys	and	cys
cys		+		+		cys
+		cys		cys		+

If recombination between the centromere and *cys2* does not occur, then the alleles will show first-division segregation (M_I).

i. Given the cross *leu3* + × + *cys2*, this tetrad is classified as a non-parental ditype. For linked genes, NPDs are the result of double crossover events and since you are told to ignore multiple crossovers, the expected frequency for this would be 0 percent.

ii. This tetrad is a parental ditype (PD). Due to random alignment and segregation during meiosis I, two linear tetrads, both classified as PD, are equally likely:

l +	and	+ c
l +		+ c
+ c		l +
+ c		l +

Therefore, you would expect 100 percent (M_I segregation of *leu3*) × 84 percent (M_I segregation of *cys2* × ½ = 42 percent of this class of tetrad.

iii. This tetrad is a tetratype (T). It shows one of the two M_I patterns for *leu3* and one of the four M_{II} patterns for *cys2*. You expect 50 percent × 16/4 percent = 2 percent of this class of tetrad.

iv. – **vii.** 0 percent. In all these tetrads, *leu3* is shown to have a second-division segregation pattern, and you are told that it always segregates at the first division.

32. a. Yes, the data indicate that the three genes are linked. The most common classes of progeny, albino (*al b⁺ fu⁺*) and brown, fuzzy (*al⁺ b fu*) represent the "parental" chromosomes. (Gene order is not specified.) The least common classes of progeny, brown (*al⁺ b fu⁺*) and albino, fuzzy (*al b⁺ fu*) represent the outcomes of double crossover events and can be used to deduce gene order. In comparing the most common to the least common, the gene "in the middle" can be determined to be *fu*. (Compare *al b⁺ fu⁺* to *al b⁺ fu* and *al⁺ b fu* to *al⁺ b fu⁺*.) To calculate map distances, you must now determine the various recombinant classes.

		al – fu	*fu – b*
710	*al fu⁺ b⁺*	P	P
698	*al⁺ fu b*	P	P
170	*al⁺ fu⁺ b⁺*	R	P
150	*al fu b*	R	P
42	*al⁺ fu b⁺*	P	R
38	*al fu⁺ b*	P	R
5	*al⁺ fu⁺ b*	R	R
3	*al fu b⁺*	R	R

So there are 170 + 150 + 5 + 3 = 328 recombinants between *al* and *fu* for a map distance of (328/progeny) × 100% = 18.1 cM and 42 + 38 + 5 + 3 = 88 recombinants between *fu* and *b* for a map distance of 4.8 cM.

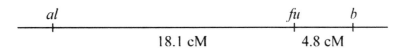

b. The most common classes of progeny (the parentals) tell you the original genotypes.

$$al\ fu^+\ b^+\ /\ al\ fu^+\ b^+\ \times\ al^+\ fu\ b\ /\ al^+\ fu\ b$$

33. a. The cross was *pro + × + his*. This makes the first tetrad class NPD (6 of these), the second tetrad class T (82 of these) and the third tetrad class PD (112 of these). When PD >> NPD, you know the two genes are linked.

b. Map distance can be calculated using the formula RF = (NPD + ½ T)100%. In this case, the frequency of NPD is 6/200 or 3 percent, and the frequency of T is 82/200 or 41 percent. Map distance is therefore 23.5 cM between these two loci.

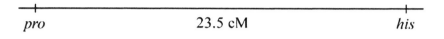

pro   23.5 cM   his

    c. To correct for multiple crossovers, the Perkins formula may be used. Thus, map distance = (T + 6NPD)50% or (0.41 + 0.18)50% = 29.5 cM.

34. **a.** The cross is t + × + m. Knowing this, you can then determine that column one (260) is PD, column two (76) is T, column three (4) is PD, column four (54) is T, column five (1) is NPD, and column six (5) is T. Given that PD >> NPD, you know the two genes are linked. For determining the distance of a gene from its centromere, you need to determine what percentage of asci show M_{II} segregation for that gene and then divide that in half to find map distance. For gene t, columns three, four, and six show M_{II} segregation (63/400) and for gene m, columns two, three, and six show M_{II} segregation (85/400). Therefore, gene t is 7.88 cM and gene m is 10.63 cM from their centromere. What remains to be determined is whether the two genes reside on the same side of their centromere or on opposite sides. The map distance between the two genes will determine which option is correct. For simple map distance, the formula to use is RF = ½T + NPD, so (½(135) + 1)/400 or 17.13 cM. This best fits the following map:

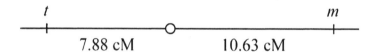

t m

7.88 cM   10.63 cM

    b. When two genes are linked, it requires a four-stranded double crossover to generate an NPD type ascus. Also, since neither gene shows M_{II} segregation, both crossovers must have occurred on the same side of the centromere.

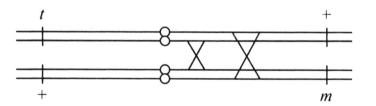

t +

+ m

35. **a.** A diagram of the *AA LL dd* parent's chromosomes would be

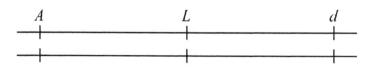

A L d

A diagram of the *aa ll dd* parent's chromosomes would be

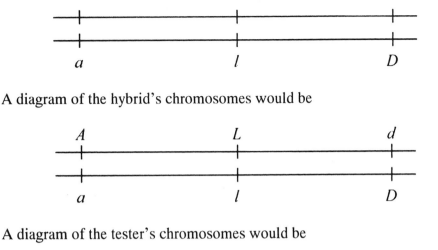

A diagram of the hybrid's chromosomes would be

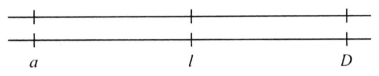

A diagram of the tester's chromosomes would be

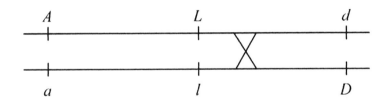

b. A diagram of the crossover(s) necessary to produce the desired genotype would be

Please note, this diagram shows a location of crossover that would give the desired genotype but does not take into consideration that each chromosome would actually consist of two chromatids.

c. The genotype of the desired crossover product is *A L D*. This is a result of a single crossover event between the *L* and *D* genes. The map distance between these genes includes both these single crossover events and also any double crossover events (simultaneous events between genes *A* and *L* and between genes *L* and *D*). The double crossovers will not give the desired outcome and therefore must be subtracted to calculate the percentage of test cross progeny with the desired phenotype. The expected double crossover is simply the chance of recombination in one interval multiplied by the chance of recombination in another interval, or in this case $0.16 \times 0.24 = 0.038$. Therefore the 24 percent recombination between genes *L* and *D* is the sum of 3.8 percent double crossover events and 20.2 percent single crossover events. Thus, the percentage of *A L D/a l d* progeny from this test cross should be $\frac{1}{2} \times 20.2 = 10.1$ percent of the total.

d. To calculate the frequency of double crossovers, we assumed that the two recombinations were independent and that there was no interference.

36. a.   The F$_1$ would have all the dominant phenotypes: presence of trichomes, tall, waxy cuticle, and presence of purple pigment.

b.   A sketch of the parents' chromosomes would be

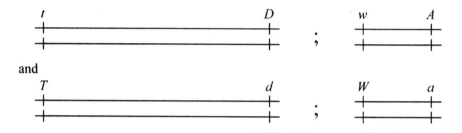

And a sketch of the F$_1$'s chromosomes would be

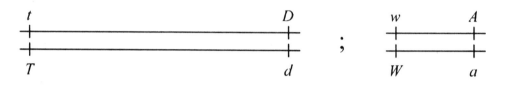

c.   For the progeny of a test cross to have all four recessive phenotypes, it must inherit a recombinant *t d* chromosome 1 and a *w a* recombinant chromosome 2 from its F$_1$ parent. Since the *T/t* and *D/d* loci are 26 m.u. apart, 26 percent of the progeny are expected to be recombinant for these genes and of these, half, or 13 percent, will be *t d*. Similarly, 8 percent of the progeny are expected to be recombinant for the *W/w* and *A/a* loci and half of these, or 4 percent, will be *w a*. Since these are independent events, 13 percent × 4 percent = 0.52 percent of the total progeny will have all four recessive phenotypes.

37. a.   The cross is *W e F / W e F* × *w E f / w E f* and the F$_1$ are *W e F / w E f*. Progeny that are *ww ee ff* from a test cross of this F$_1$ must have inherited one of the double crossover recombinant chromosomes (*w e f*). Assuming no interference, the expected percentage of double crossovers is 8 percent × 24 percent = 1.92 percent and half of this is 0.96 percent.

b.   To obtain a *ww ee ff* progeny from a self cross of this F$_1$ requires the independent inheritance of two doubly recombinant *w e f* chromosomes. The chance of this, based on the answer to part (a) of this question would be 0.96 × 0.96 = 0.009 percent.

38.   Since the cross is + + × c m, the first class of tetrad is PD, the second class is T, and the third class is NPD. That PD >> NPD tells you the genes are linked.

a.   RF = (½T + NPD)100% = (0.205 + 0.03)100% = 23.5%

b.   The Perkins' formula is RF = (T + 6 NPD)50% = 29.5%

The formula for the Haldane mapping function is RF = ½(1 − e^{-m}).
Solving for this situation, e^{-m} = 1 − (2 × 0.235) = 0.53 so m = 0.635.

To convert to map distance, m, the measure of crossover frequency is
multiplied by 50 percent, to give a corrected map distance of 31.74%.

c.   Yes. The Perkins' formula only takes into account 0, 1, and 2 crossovers
while the Haldane mapping formula has no limitations on the number of
crossover events. It predicts the chance of any nonzero number of
crossovers.

39.   There are three patterns of possible outcomes from a test cross of a triply
heterozygous parent. If all genes are linked, you expect pairs of roughly equal
numbers of progeny in four different frequencies (for example, see problem 20
of this chapter). If all genes are unlinked, the expectation is eight classes of
progeny in roughly equal numbers due to the independent assortment of all
genes. The last possibility is that two of the genes are linked but the third is
unlinked. In that case, the expectation is two groups of four of two different
frequencies. This final pattern is observed in the data of this problem.

In reviewing the data, the four most common classes are either waltzing bent
or nonwaltzing straight with gray or albino segregating equally among those
two groups. This observation tells you that the *W/w* and *B/b* genes are linked
and both are unlinked to the *G/g* gene.

a.   *W/w • G/G • A/A × w/w • g/g • a/a*

b.

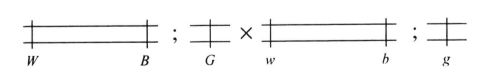

c.   For this cross, *W B* and *w b* begin linked to each other so any progeny of
the test cross that are *W b* or *w B* are recombinant.

map distance = (4 + 5 + 5 + 6)/100 × 100% = 20 map units

40.   The cross is + ; + × *f* ; *p* and you know that *f* is linked to its centromere and
only segregates at the first division (M$_I$). The *p* gene, on average, segregates at
the second divisions (M$_{II}$) 12 percent of the time (and therefore segregates at
the first division 88 percent of the time).

a. All octads will show M_I segregation for the $+/f$ gene and 88 percent will show M_I segregation for the $+/p$ gene. Since the genes are unlinked, PD = NPD so half of all octads that meet these conditions will be PD, or in this case, 44 percent.

b. 44 percent (see part (a)).

c. Since all octads will show M_I segregation for $+/f$, all M_{II} patterns for $+/p$ will be tetratypes or, in this case, 12 percent.

d. 0 percent. $+/f$ does not segregate at the second division.

41. The cross is $al\text{-}2 + ; + met\text{-}l \times + arg\text{-}6 ; lys\text{-}5 +$. The transient diploid will be $al\text{-}2 +/ + arg\text{-}6 ; + met\text{-}1/lys\text{-}5 +$. From this, one half of the recombinant chromosomes 1 will be $+ +$ and one half of the recombinant chromosomes 6 will be $+ +$. Since these are independent events, $\frac{1}{2}(30 \text{ percent}) \times \frac{1}{2}(20 \text{ percent}) = 1.5$ percent of the progeny will be $+ + ; + +$.

42. **a.** The data include that the three genes are linked. The most common classes of progeny, $k c +$ and $+ + e$ represent the "parental" chromosomes. (Gene order is not specified.) The least common classes of progeny, $k c e$ and $+ + +$ represent the outcomes of double crossover events and can be used to deduce gene order. In comparing the most common to the least common, the gene "in the middle" can be determined to be e. (Compare $k c +$ to $k c e$ and $+ + e$ to $+ + +$.) To calculate map distances, you must now determine the various recombinant classes.

		$k-e$	$e-c$
899	$+ + e$	P	P
876	$k c +$	P	P
67	$k + e$	R	P
58	$+ c +$	R	P
49	$k + +$	P	R
44	$+ c e$	P	R
4	$+ + +$	R	R
3	$k c e$	R	R

So there are $67 + 58 + 4 + 3 = 132$ recombinants between k and e for a map distance of (132/total progeny) $\times 100\% = 6.6$ cM and $49 + 44 + 4 + 3 = 100$ recombinants between e and c for a map distance of 5 cM.

b.

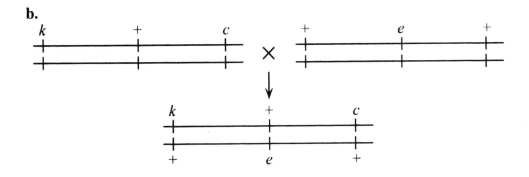

c. The expected frequency of double crossovers is 6.6 percent × 5 percent = 0.33 percent. The actual frequency of double crossovers is (4 + 3)/2000 × 100% = 0.35 percent. To calculate interference, the formula is I = 1 − (observed DCO/expected DCO) = -0.06. In this case, the interference is "negative." The occurrence of one recombination event appears to slightly increase the chances of another. Of course, the difference between observed (7) and expected (6.6) is not statistically meaningful. While there are observed instances of negative interference, generally, interference values are between 0 (no interference) and 1 (complete interference).

43. The cross is *A/A • B/B × a/a • b/b* and then the resulting F₁ dihybrid is test crossed. Therefore, for calculation purposes, test cross progeny that are *A/a • B/b* or *a/a • b/b* should be considered "parentals" and progeny that are *A/a • b/b* or *a/a • B/b* should be considered "recombinants."

The following shows the expected proportions of parental (P) and recombinant (R) genotypes for independent assortment and the three RF values for this problem.

RF

	0.5	0.3	0.2	0.1
P	0.25	0.35	0.4	0.45
P	0.25	0.35	0.4	0.45
R	0.25	0.15	0.1	0.05
R	0.25	0.15	0.1	0.05

In birth order, the progeny are

A/a • B/b parental

a/a • b/b parental

A/a • B/b parental

A/a • b/b recombinant

a/a • b/b parental

A/a • B/b parental

a/a • B/b recombinant

The probability of obtaining these results if the two genes are unlinked (RF of 50 percent) will be $0.25 \times 0.25 \times 0.25 \times 0.25 \times 0.25 \times 0.25 \times 0.25 \times B = 0.000061B$ (where B = the number of possible birth orders for five parental and two recombinant individuals).

a. The probability of obtaining these results if the two genes are linked (RF of 10 percent) will be $0.45 \times 0.45 \times 0.45 \times 0.05 \times 0.45 \times 0.45 \times 0.05 \times B = 0.000046B$. The ratio of this to the outcome expected if the genes are unlinked is $0.000046B/0.000061B = 0.756$. The logarithm of this ratio is the Lod score, or in this case, $Lod = \log(0.756) = -0.12$.

b. The probability of obtaining these results if the two genes are linked (RF of 20 percent) will be $0.40 \times 0.40 \times 0.40 \times 0.10 \times 0.40 \times 0.40 \times 0.10 \times B = 0.0000102B$. The ratio of this to the outcome expected if the genes are unlinked is 1.68 and the Lod score will be 0.22.

c. The probability of obtaining these results if the two genes are linked (RF of 30 percent) will be 0.000118B, the ratio will be 1.93 and the Lod score will be 0.29.

44.   The formula for the Haldane mapping function is $RF = \frac{1}{2}(1 - e^{-m})$. Solving for a measured RF = 5%, $e^{-m} = 1 - (2 \times 0.05) = 0.90$, so $m = 0.105$. To convert to map distance, m, the measure of crossover frequency, is multiplied by 50 percent, to give a corrected map distance of 5.27%. Similarly, an RF = 10% would be corrected to 11.2%, RF = 20% to 25.5%, RF = 30% to 46% and RF = 40% to 80%! A graph of this data is presented below.

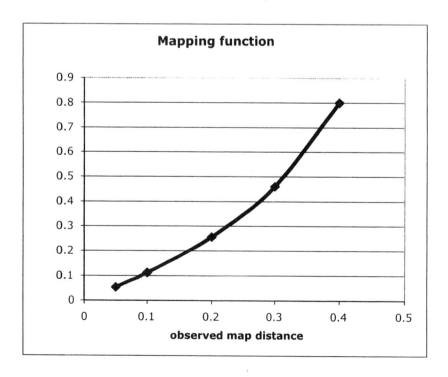

As can be observed, when map distances are small, the observed and corrected recombination frequencies are much the same. So the larger the RF, the more important it is to correct for under estimation of map distance by use of the mapping function.

CHALLENGING PROBLEMS

45. a. All of these genes are linked. To determine this, each gene pair is examined separately. For example, are *A* and *B* linked?

$A B = 140 + 305 = 445$
$a b = 145 + 310 = 455$
$a B = 42 + 6 = 48$
$A b = 43 + 9 = 52$

Conclusion: the two genes are linked and 10 m.u. apart.

Are *A* and *D* linked?

$A D = 0$
$a d = 0$
$A d = 43 + 140 + 9 + 305 = 497$
$a D = 42 + 145 + 6 + 310 = 503$

Conclusion: the two genes show no recombination and at this resolution, are 0 m.u. apart.

Are *B* and *C* linked?

$B C = 42 + 140 = 182$
$b c = 43 + 145 = 188$
$B c = 6 + 305 = 311$
$b C = 9 + 310 = 319$

Conclusion: the two genes are linked and 37 m.u. apart.

Are *C* and *D* linked?

$C D = 42 + 310 = 350$
$c d = 43 + 305 = 348$
$C d = 140 + 9 = 149$
$c D = 145 + 6 = 151$

Conclusion: the two genes are linked and 30 m.u. apart. Therefore, all four genes are linked.

b. and c. Because *A* and *D* show no recombination, first rewrite the progeny omitting *D* and *d* (or omitting *A* and *a*).

a B C	42
A b c	43
A B C	140
a b c	145
a B c	6
A b C	9
A B c	305
a b C	310
	1000

Note that the progeny now look like those of a typical three point test cross, with *A B c* and *a b C* the parental types (most frequent) and *a B c* and *A b C* the double recombinants (least frequent). The gene order is *B A C*. This is determined either by the map distances or by comparing double recombinants with the parentals; the gene that switches in reference with the other two is the gene in the center (*B A c* → *B a c*, *b a C* → *b A C*).

Next, rewrite the progeny again, this time putting the genes in the proper order, and classify the progeny.

B a C	42	CO *A–B*
b A c	43	CO *A–B*
B A C	140	CO *A–C*
b a c	145	CO *A–C*
B a c	6	DCO
b A C	9	DCO
B A c	305	parental
b a C	310	parental

To construct the map of these genes, use the following formula:

distance between two genes = $\dfrac{(100\%) \text{ (number of single CO + number of DCO)}}{\text{total number of progeny}}$

For the *A* to *B* distance

$$= \frac{(100\%)(42 + 43 + 6 + 9)}{1000} = 10 \text{ m.u.}$$

For the *A* to *C* distance

$$= \frac{(100\%)(140 + 145 + 6 + 9)}{1000} = 30 \text{ m.u.}$$

The map is

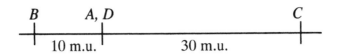

The parental chromosomes actually were *B (A,d) c/b (a,D) C*, where the parentheses indicate that the order of the genes within is unknown.

   d.   Interference = 1 – (observed DCO/expected DCO)

$$= 1 - (6 + 9)/[(0.10)(0.30)(1000)]$$
$$= 1 - 15/30 = 0.5$$

46.   The verbal description indicates the following cross and result:

P   *N/– · A/– × n/n · O/O*

F$_1$ *N/n · A/O × N/n · A/O*

The results indicate linkage, so the cross and results can be rewritten:
P   *N A/– – × n O/n O*

F$_1$ *N A/ n O × N A/n O*

F$_2$ 66%   *N A/– – or N –/– A*
            16%   *n O/n O*
             9%   *n A/n –*
             9%   *N O/– O*

Only one genotype is fully known: 16 percent *n O/n O*, a combination of two parental gametes. The frequency of two parental gametes coming together is the frequency of the first times the frequency of the second. Therefore, the frequency of each *n O* gamete is the square root of 0.16, or 0.4. Within an organism the two parental gametes occur in equal frequency. Therefore, the frequency of *N A* is also 0.4. The parental total is 0.8, leaving 0.2 for all recombinants. Therefore, *N O* and *n A* occur at a frequency of 0.1 each. The two genes are 20 m.u. apart. Complete frequencies for all genotypes contributing to the four phenotypes can be obtained from a punnett square using the gamete frequencies provided above.

47.   **a. and b.** The data indicate that the progeny males have a different phenotype than the females. Therefore, all the genes are on the X chromosome. The two most frequent phenotypes in the males indicate the genotypes of the X chromosomes in the female, and the two least frequent phenotypes in the males indicate the gene order. Gene z is in the middle. Data from only the males are used to determine map distances. The cross is:

P $x\, z\, y^+/x^+\, z^+\, y \times x^+\, z^+\, y^+/Y$

F$_1$ males

430	$x\, z\, y^+/Y$	parental
441	$x^+\, z^+\, y/Y$	parental
39	$x\, z\, y/Y$ CO z–y	
30	$x^+\, z^+\, y^+/Y$	CO z–y
32	$x^+\, z\, y^+/Y$	CO x–z
27	$x\, z^+\, y/Y$	CO x–z
1	$x^+\, z\, y/Y$	DCO
0	$x\, z^+\, y^+/Y$	DCO

c. z–y: $100\%(39 + 30 + 1)/1000 = 7.0$ m.u.
     x–z: $100\%(32 + 27 + 1)/1000 = 6.0$ m.u.

c.c.   = observed DCO/expected DCO
       = $1/[(0.06)(0.07)(1000)] = 0.238$

48. The data given for each of the three-point test crosses can be used to determine the gene order by realizing that the rarest recombinant classes are the result of double cross-over events. By comparing these chromosomes to the "parental" types, the alleles that have switched represent the gene in the middle.

For example, in (1), the most common phenotypes (+ + + and $a\, b\, c$) represent the parental allele combinations. Comparing these to the rarest phenotypes of this data set (+ $b\, c$ and a + +) indicates that the a gene is recombinant and must be in the middle. The gene order is $b\, a\, c$.

For (2), + $b\, c$ and a + + (the parentals) should be compared to + + + and $a\, b\, c$ (the rarest recombinants) to indicate that the a gene is in the middle. The gene order is $b\, a\, c$.

For (3), compare + b + and a + c with $a\, b$ + and + + c, which gives the gene order $b\, a\, c$.

For (4), compare + + c and $a\, b$ + with + + + and $a\, b\, c$, which gives the gene order $a\, c\, b$.

For (5), compare + + + and $a\, b\, c$ with + + c and $a\, b$ +, which gives the gene order $a\, c\, b$.

49. The gene order is $a\, c\, b\, d$.

Recombination between *a* and *c* occurred at a frequency of

100%(139 + 3 + 121 + 2)/(669 + 139 + 3 + 121 + 2 + 2,280 + 653 + 2,215)
= 100%(265/6,082) = 4.36%

Recombination between *b* and *c* in cross 1 occurred at a frequency of

100%(669 + 3 + 2 + 653)/(669 + 139 + 3 + 121 + 2 + 2,280 + 653 + 2,215)
= 100%(1,327/6,082) = 21.82%

Recombination between *b* and *c* in cross 2 occurred at a frequency of
100%(8 + 14 + 153 + 141)/(8 + 441 + 90 + 376 + 14 + 153 + 65 + 141)
= 100%(316/1,288) = 24.55%

The difference between the two calculated distances between *b* and *c* is not surprising because each set of data would not be expected to yield exactly identical results. Also, many more offspring were analyzed in cross 1. Combined, the distance would be

100%[(316 + 1,327)/(1,288 + 6,082)] = 22.3%

Recombination between *b* and *d* occurred at a frequency of
100%(8 + 90 + 14 + 65)/(8 + 441 + 90 + 376 + 14 + 153 + 65 + 141)
= 100%(177/1,288) = 13.68%

50. Part (a) of this problem is solved two ways, once in the standard way, once in a way that emphasizes a more mathematical approach.

The cross is

P   *P A R/P A R* × *p a r/p a r*

F$_1$ *P A R/p a r* × *p a r/p a r*, a three-point test cross

a. In order to find what proportion will have the Vulcan phenotype for all three characteristics, we must determine the frequency of parentals. Crossing-over occurs 15 percent of the time between P and A, which means it does not occur 85 percent of the time. Crossing-over occurs 20 percent of the time between A and R, which means that it does not occur 80 percent of the time.

p (no crossover between either gene)
= *p*(no crossover between *P* and *A*) × *p*(no crossover between A and *R*)
= (0.85)(0.80) = 0.68

Half the parentals are Vulcan, so the proportion that are completely Vulcan is $^1/_2$(0.68) = 0.34

Mathematical method

Number of parentals = 1 – (single CO individuals – DCO individuals)
= 1 – {[0.15 + 0.20 – 2(0.15)(0.20)] – (0.15) (0.20)} = 0.68

Because half the parentals are Earth alleles and half are Vulcan, the frequency of children with all three Vulcan characteristics is $1/2(0.68)$ = 0.34

b.   Same as above, 0.34

c.   To yield Vulcan ears and hearts and Earth adrenals, a crossover must occur in both regions, producing double crossovers. The frequency of Vulcan ears and hearts and Earth adrenals will be half the DCOs, or $1/2(0.15) (0.20)$ = 0.015

d.   To yield Vulcan ears and an Earth heart and adrenals, a single crossover must occur between *P* and *A*, and no crossover can occur between *A* and *R*. The frequency will be

p(CO P–A) × p(no CO A–R) = (0.15)(0.80) = 0.12

Of these, $1/2$ are *P a r* and $1/2$ are *p A R*. Therefore, the proportion with Vulcan cars and an Earth heart and adrenals is 0.06

51. **a.**   To obtain a plant that is *a b c/a b c* from selfing of *A b c/a B C*, both gametes must be derived from a crossover between *A* and *B*. The frequency of the *a b c* gamete is

$1/2 \, p$(CO A–B) × p(no CO B–C) = $1/2(0.20)(0.70)$ = 0.07

Therefore, the frequency of the homozygous plant will be $(0.07)^2$ = 0.0049

b.   The cross is *A b c/a B C × a b c/a b c*.

To calculate the progeny frequencies, note that the parentals are equal to all those that did not experience a crossover. Mathematically this can be stated as

parentals   = p(no CO A–B) × p(no CO B–C)
= (0.80)(0.70) = 0.56

Because each parental should be represented equally

A b c = $1/2(0.56)$ = 0.28

$a\,B\,C = \frac{1}{2}(0.56) = 0.28$

As calculated above, the frequency of the *a b c* gamete is

$\frac{1}{2}\,p(\text{CO }A\text{–}B) \times p(\text{no CO }B\text{–}C) = \frac{1}{2}(0.20)(0.70) = 0.07$

as is the frequency of *A B C*.

The frequency of the *A b C* gamete is

$\frac{1}{2}\,p(\text{CO }B\text{–}C) \times p(\text{no CO }A\text{–}B) = \frac{1}{2}(0.30)(0.80) = 0.12$

as is the frequency of *a B c*.

Finally, the frequency of the *A B c* gamete is

$\frac{1}{2}\,p(\text{CO }A\text{–}B) \times p(\text{CO }B\text{–}C) = \frac{1}{2}(0.20)(0.30) = 0.03$

as is the frequency of *a b C*.

So for 1,000 progeny, the expected results are

A b c	280
a B C	280
A B C	70
a b c	70
A b C	120
a B c	120
A B c	30
a b C	30

c.   Interference = 1 – observed DCO/expected DCO
     0.2 = 1 – observed DCO/(0.20)(0.30)

observed DCO = (0.20)(0.30) – (0.20)(0.20)(0.30) = 0.048

The *A–B* distance = 20% = 100% [*p*(CO *A–B*) + *p*(DCO)].
Therefore, *p*(CO *A–B*) = 0.20 – 0.048 = 0.152

Similarly, the *B–C* distance = 30% = 100% [*p*(CO *B–C*) + *p*(DCO)]
Therefore, *p*(CO *B–C*) = 0.30 – 0.048 = 0.252

The *p*(parental) = 1 – *p*(CO *A–B*) – *p*(CO *B–C*) – *p*(observed DCO)
     = 1 – 0.152 – 0.252 – 0.048 = 0.548

So for 1,000 progeny, the expected results are

A b c	274
a B C	274
A B C	76
a b c	76
A b C	126
a B c	126
A B c	24
a b C	24

52. **a.** Blue sclerotic (*B*) appears to be an autosomal dominant disorder. Hemophilia (*h*) appears to be an X-linked recessive disorder.

b. If the individuals in the pedigree are numbered as generations I through IV and the individuals in each generation are numbered clockwise, starting from the top right-hand portion of the pedigree, their genotypes are

I: *b/b* ; *H/h, B/b* ; *H/*Y

II: *B/b* ; *H/*Y, *B/b* ; *H/*Y, *b/b* ; *H/*Y, *B/b* ; *H/h, b/b* ; *H/*Y, *B/b* ; *H/h, B/b* ; *H/h, B/b* ; *H/–, b/b* ; *H/–*

III: *b/b* ; *H/–, B/b* ; *H/–, b/b* ; *h/*Y, *b/b* ; *H/*Y, *B/b* ; *H/*Y, *B/b* ; *H/–, B/b* ; *H/*Y, *B/b* ; *h/*Y, *B/b* ; *H/–, b/b* ; *H/*Y, *B/b* ; *H/–, b/b* ; *H/*Y, *B/b* ; *H/–, B/b* ; *H/*Y, *B/b* ; *h/*Y, *b/b* ; *H/*Y, *b/b* ; *H/*Y, *b/b* ; *H/–, b/b* ; *H/*Y, *b/b* ; *H/*Y, *B/b* ; *H/–, B/b* ; *H/*Y, *B/b* ;*h/*Y

IV: *b/b* ; *H/–, B/b* ; *H/–, B/b* ; *H/–, b/b* ; *H/h, b/b* ; *H/h, b/b* ; *H/*Y, *b/b* ; *H/H, b/b* ; *H/*Y, *b/b* ; *H/h, b/b* ; *H/H, b/b* ; *H/H, b/b* ; *H/*Y, *b/b* ; *H/*Y, *b/b* ; *H/H, b/b* ; *H/*Y, *b/b* ; *H/*Y, *B/b* ; *H/*Y, *b/b* ; *H/*Y, *b/b* ; *H/–, b/b* ; *H/*Y, *b/b* ; *H/*Y, *b/b* ; *H/–, b/b* ; *H/H, b/b* ; *H/–, b/b* ; *H/–, b/b* ; *H/*Y, *b/b* ; *H/*Y, *b/b* ; *H/*Y, *b/b* ; *H/h, B/b* ; *H/h, B/b* ; *H/*Y, *b/b* ; *H/*Y, *B/b* ; *H/*Y, *b/b* ; *H/h*

c. There is no evidence of linkage between these two disorders. Because of the modes of inheritance for these two genes, no linkage would be expected.

d. The two genes exhibit independent assortment.

e. No individual could be considered intrachromosomally recombinant. However, a number show interchromosomal recombination, for example, all individuals in generation III that have both disorders.

53. If *h* = hemophilia and *b* = colorblindness, the genotypes for individuals in the pedigree can be written as

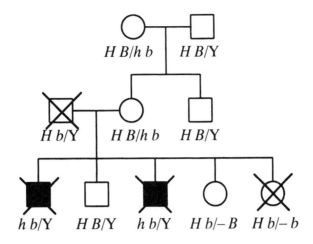

The mother of the two women in question would produce the following gametes:

0.45   *H B*
0.45   *h b*
0.05   *h B*
0.05   *H b*

Woman III-4 can be either *H b/H B* (0.45 chance) or *H b/h B* (0.05 chance), because she received *B* from her mother. If she is *H b/h B* [0.05/(0.45 + 0.05) = 0.10 chance], she will produce the parental and recombinant gametes with the same probabilities as her mother. Thus, her child has a 45 percent chance of receiving *h B*, a 5 percent chance of receiving *h b*, and a 50 percent chance of receiving a Y from his father. The probability that her child will be a hemophiliac son is (0.1)(0.5)(0.5) = 0.025 = 2.5 percent.

Woman III-5 can be either *H b/H b* (0.05 chance) or *H b/h b* (0.45 chance), because she received *b* from her mother. If she is *H b/h b* [0.45/(0.45 + 0.05) = 0.90 chance], she has a 50 percent chance of passing *h* to her child, and there is a 50 percent chance that the child will be male. The probability that she will have a son with hemophilia is (0.9)(0.5)(0.5) = 0.225 = 22.5 percent.

54. a.   Cross 1 reduces to

P $A/A \cdot B/B \cdot D/D \times a/a \cdot b/b \cdot d/d$

F_1 $A/a \cdot B/b \cdot D/d \times a/a \cdot b/b \cdot d/d$

The test cross progeny indicate these three genes are linked.

Test cross	A B D	316	parental
progeny	a b d	314	parental
	A B d	31	CO *B–D*
	a b D	39	CO *B–D*

A b d	130	CO *A–B*
a B D	140	CO *A–B*
A b D	17	DCO
a B d	13	DCO

A–B: 100%(130 + 140 + 17 + 13)/1000 = 30 m.u.
B–D: 100%(31 + 39 + 17 + 13)/1000 = 10 m.u.

Cross 2 reduces to

P $A/A \cdot C/C \cdot E/E \times a/a \cdot c/c \cdot e/e$

F_1 $A/a \cdot C/c \cdot E/e \times a/a \cdot c/c \cdot e/e$

The test cross progeny indicate these three genes are linked.

Test cross	*A C E*	243	parental
progeny	*a c e*	237	parental
	A c e	62	CO *A–C*
	a C E	58	CO *A–C*
	A C e	155	CO *C–E*
	a c E	165	CO *C–E*
	a C e	46	DCO
	A c E	34	DCO

A–C: 100% (62 + 58 + 46 + 34)/1000 = 20 m.u.
C–E: 100% (155 + 165 + 46 + 34)/1000 = 40 m.u.

The map that accommodates all the data is

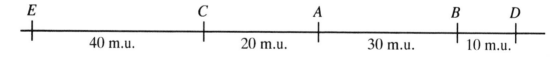

E		C		A		B	D
	40 m.u.		20 m.u.		30 m.u.	10 m.u.	

b.   Interference (I) = 1 – [(observed DCO)/(expected DCO)]

For cross 1: I = 1 – {30/[(0.30)(0.10)(1000)]} = 1 – 1 = 0, no interference

For cross 2: I = 1 – {80/[(0.20)(0.40)(1000)]} = 1 – 1 = 0, no interference

55. a.   The first F_1 is *L H/l h* and the second is *l H/L h*. For progeny that are *l h/l h*, they have received a "parental" chromosome from the first F_1 and a "recombinant" chromosome from the second F_1. The genes are 16 percent apart so the chance of a parental chromosome is $1/2$(100 – 16%) = 42% and the chance of a recombinant chromosome is $1/2$(16%) = 8%.

The chance of both events = 42% × 8% = 3.36%

b. To obtain *Lh*/*l h* progeny, either a parental chromosome from each parent was inherited *or* a recombinant chromosome from each parent was inherited. The total probability will therefore be

(42% × 42%) + (8% × 8%) = (17.6% + 0.6%) = 18.2%

56. Crossing-over occurs 8 percent of the time between *w* and *s*, which means it does not occur 92 percent of the time. Crossing-over occurs 14 percent of the time between *s* and *e*, which means that it does not occur 86 percent of the time.

a. and b. The frequency of parentals = *p* (no crossover between either gene)

$$= p(\text{no CO } w\text{–}s) \times p(\text{no CO } s\text{–}e) = (0.92)(0.86)$$
$$= 0.791$$

or

$$\tfrac{1}{2}(0.791) = 0.396 \text{ each}$$

c. and d. The frequency that will show recombination between *w* and *s* only

$$= p(\text{CO } w\text{–}s) \times p(\text{no CO } s\text{–}e) = (0.08)(0.86) = 0.069$$

or

$$\tfrac{1}{2}(0.069) = 0.035 \text{ each}$$

e. and f. The frequency that will show recombination between *s* and *e* only

$$= p(\text{CO } s\text{–}e) \times p(\text{no CO } w\text{–}s) = (0.14)(0.92) = 0.128$$

or

$$\tfrac{1}{2}(0.128) = 0.064 \text{ each}$$

g. and h. The frequency that will show recombination between *w* and *s* and *s* and *e*

$$= p(\text{CO } w\text{–}s) \times p(\text{CO } s\text{–}e) = (0.08)(0.14) = 0.011$$

or

$$\tfrac{1}{2}(0.011) = 0.006 \text{ each}$$

57. This problem is analogous to meiosis in organisms that form linear tetrads. Let red = *R* and blue = *r*. This can now be compared to meiosis in an organism that is *R*/*r*. The patterns, their frequencies, and the division of segregation are given below. Notice that the probabilities change as each ball/allele is selected. This occurs when there is sampling without replacement.

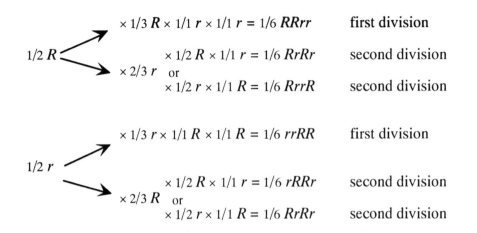

$\times\ 1/3\ R \times 1/1\ r \times 1/1\ r = 1/6\ \boldsymbol{RRrr}$ first division

$\times\ 1/2\ R \times 1/1\ r = 1/6\ \boldsymbol{RrRr}$ second division

$\times\ 1/3\ r \times 1/1\ R \times 1/1\ R = 1/6\ \boldsymbol{rrRR}$ first division

$\times\ 1/2\ R \times 1/1\ r = 1/6\ \boldsymbol{rRRr}$ second division

These results indicate one-third first-division segregation and two-thirds second-division segregation.

58. As the problem suggests, calculate the frequencies of the various possibilities. The percentage of tetrads without crossing over is 88% × 80% = 70.4%. The percentage of tetrads with a single crossover in region (i) and none in region (ii) is 12% × 80% = 9.6%. The percentage of tetrads with a single crossover in region (ii) and none in region (i) is 20% × 88% = 17.6% and the percentage of tetrads with crossovers in both regions is 12% × 20% = 2.4%.

Now work out the patterns of segregation that result in each case

For no crossovers

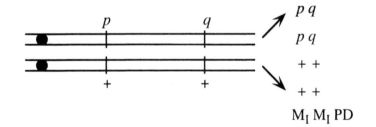

For a single crossover in region (i)

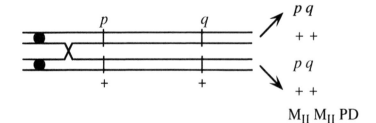

For a single crossover in region (ii)

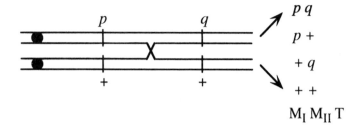

For double crossovers, there are four types, all equally likely

two-strand

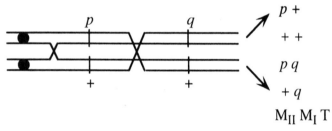

four-strand

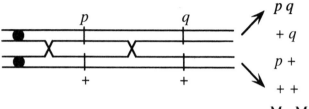

and two different three-strand

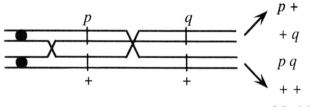

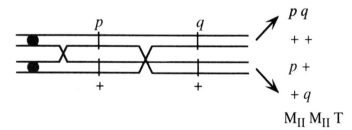

(a) M_I M_I PD is the result of no crossovers = 70.4%

(b) M_I M_I NPD is not found as a result

(c) M_I M_{II} T is the result of a single crossover in region (ii) = 17.6%

(d) M_{II} M_I T is the result of the two- and four-strand double crossovers = 1.2%

(e) M_{II} M_{II} PD is the result of a single crossover in region (i) = 9.6%

(f) M_{II} M_{II} NPD is not found as result

(g) M_{II} M_{II} T is the result of both three-strand double crossovers = 1.2%

59. **a. and b.** The data support the independent assortment of two genes (call them *arg1* and *arg2*). The cross becomes *arg1* ; *arg2⁺* × *arg1⁺* ; *arg2* and the resulting tetrads are:

4 : 0 (PD)	3 : 1 (T)	2 : 2 (NPD)
arg1 ; *arg2⁺*	*arg1* ; *arg2⁺*	*arg1* ; *arg2*
arg1 ; *arg2⁺*	*arg1⁺* ; *arg2*	*arg1* ; *arg2*
arg1⁺ ; *arg2*	*arg1* ; *arg2*	*arg1⁺* ; *arg2⁺*
arg1⁺ ; *arg2*	*arg1⁺* ; *arg2⁺*	*arg1⁺* ; *arg2⁺*

Because PD = NPD, the genes are unlinked.

60. **a.** The cross is $A/A \bullet B/B \bullet l/l \times a/a \bullet b/b \bullet s/s$. The resulting F_1 is backcrossed to the second parent and the backcross progeny indicate that all three markers are linked.

b. and c. By comparing the most common progeny to the least (*A B l/a b s* to *A B s/a b s* and *a b s/abs* to *a b l/a b s*) we can deduce that the RFLP allele is between genes *A/a* and *B/b*. The distance between these markers is (43 + 37 + 9 + 11)/1000 = 10 map units between gene *A/a* and the RFLP and (93 + 87 + 9 + 11)/1000 = 20 map units between the RFLP and gene *B/b*. The restriction sites are not drawn to scale relative to the genetic distances.

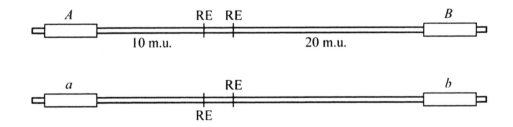

5

The Genetics of Bacteria and Their Viruses

BASIC PROBLEMS

1. An Hfr strain has the fertility factor F integrated into the chromosome. An F^+ strain has the fertility factor free in the cytoplasm. An F^- strain lacks the fertility factor.

2. All cultures of F^+ strains have a small proportion of cells in which the F factor is integrated into the bacterial chromosome and are, by definition, Hfr cells. These Hfr cells transfer markers from the host chromosome to a recipient during conjugation.

3. **a.** Hfr cells involved in conjugation transfer host genes in a linear fashion. The genes transferred depend on both the Hfr strain and the length of time during which the transfer occurred. Therefore, a population containing several different Hfr strains will appear to have an almost random transfer of host genes. This is similar to generalized transduction, in which the viral protein coat forms around a specific amount of DNA rather than specific genes. In generalized transduction, any gene can be transferred.

    b. F' factors arise from improper excision of an Hfr from the bacterial chromosome. They can have only specific bacterial genes on them because the integration site is fixed for each strain. Specialized transduction resembles this in that the viral particle integrates into a specific region of the bacterial chromosome and then, upon improper excision, can take with it only specific bacterial genes. In both cases, the transferred gene exists as a second copy.

4. Generalized transduction occurs with lytic phages that enter a bacterial cell, fragment the bacterial chromosome, and then, while new viral particles are being assembled, improperly incorporate some bacterial DNA within the viral protein coat. Because the amount of DNA, not the information content of the

DNA, is what governs viral particle formation, any bacterial gene can be included within the newly formed virus. In contrast, specialized transduction occurs with improper excision of viral DNA from the host chromosome in lysogenic phages. Because the integration site is fixed, only those bacterial genes very close to the integration site will be included in a newly formed virus.

5. While the interrupted-mating experiments will yield the gene order, it will be relative only to fairly distant markers. Thus, the precise location cannot be pinpointed with this technique. Generalized transduction will yield information with regard to very close markers, which makes it a poor choice for the initial experiments because of the massive amount of screening that would have to be done. Together, the two techniques allow, first, for a localization of the mutant (interrupted-mating) and, second, for precise determination of the location of the mutant (generalized transduction) within the general region.

6. This problem is analogous to forming long gene maps with a series of three-point testcrosses. Arrange the four sequences so that their regions of overlap are aligned:

$$\overline{M-Z}-X-W-C$$
$$W-C-N-A-L$$
$$A-L-B-R-U$$
$$B-R-U-\underline{M-Z}$$

The regions with the bars above or below are identical in sequence (and "close" the circular chromosome). The correct order of markers on this circular map is

$$-M-Z-X-W-C-N-A-L-B-R-U-$$

7. First, carry out a series of crosses in which you select in a long mating each of the auxotrophic markers. Thus, select for arg^+ T^r. In each case score for penicillin resistance. Although not too informative, these crosses will give the marker that is closest to pen^r by showing which marker has the highest linkage. Then, do a second cross concentrating on the two markers on either side of the pen^r locus. Suppose that the markers are *ala* and *glu*. You can first verify the order by taking the cross in which you selected for ala^+, the first entering marker, and scoring the percentage of both pen^r and glu^+. Because of the gradient of transfer, the percentage of pen^r should be higher than the percentage of glu^+ among the selected ala^+ recombinants.

Then, take the mating in which glu^+ was the selected marker. Because this marker enters last, one can use the cross data to determine the map units by determining the percentage of colonies that are ala^+ pen^r, and by the number of ala^- pen^r colonies.

8.   **a.**   Determine the gene order by comparing *arg⁺ bio⁺ leu⁻* with *arg⁺ bio⁻ leu⁺*. If the order were *arg leu bio*, four crossovers would be required to get *arg⁺ leu⁻ bio⁺*, while only two would be required to get *arg⁺ leu⁺ bio⁻*. If the order is *arg bio leu*, four crossovers would be required to get *arg⁺ bio⁻ leu⁺*, and only two would be required to get *arg⁺ bio⁺ leu⁻*. There are eight recombinants that are *arg⁺ bio⁺ leu⁻* and none that are *arg⁺ bio⁻ leu⁺*. On the basis of the frequencies of these two classes, the gene order is *arg bio leu*.

   b.   The *arg-bio* distance is determined by calculating the percentage of the exconjugants that are *arg⁺ bio⁻ leu⁻*. These cells would have had a crossing-over event between the *arg* and *bio* genes.
   RF = 100%(48)/376 = 12.77 m.u.

   Similarly, the *bio-leu* distance is estimated by the *arg⁺ bio⁺ leu⁻* colony type.
   RF = 100%(8)/376 = 2.13 m.u.

9.   The most straightforward way would be to pick two Hfr strains that are near the genes in question but are oriented in opposite directions. Then, measure the time of transfer between two specific genes, in one case when they are transferred early and in the other when they are transferred late. For example,

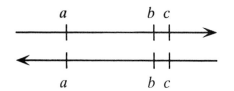

10.   The best explanation is that the integrated F factor of the Hfr looped out of the bacterial chromosome abnormally and is now an F′ that contains the *pro⁺* gene. This F′ is rapidly transferred to F⁻ cells, converting them to *pro⁺* (and F⁺).

11.   The high rate of integration and the preference for the same site originally occupied by the F factor suggest that the F′ contains some homology with the original site. The source of homology could be a fragment of the F factor, or more likely, it is homology with the chromosomal copy of the bacterial gene that is also present on the F′.

12.   First, carry out a cross between the Hfr and F⁻, and then select for colonies that are *ala⁺ strʳ*. If the Hfr donates the *ala* region late, then redo the cross but now interrupt the mating early and select for *ala⁺*. This selects for an F′, because this Hfr would not have transferred the *ala* gene early.

   If the Hfr instead donates this region early, then use a Rec⁻ strain that cannot incorporate a fragment of the donor chromosome by recombination. Any *ala⁺*

colonies from the cross should then be used in a second mating to another *ala*⁻ strain to see whether they can donate the *ala* gene easily, which would indicate that there is F′ *ala*. (This would also require another marker to differentiate the donor and recipient strains. For example, the *ala*⁻ strain could be tetracycline^r and selection would be for *ala*$^+$ *tet*^{r}.)

13. **a. and b.**

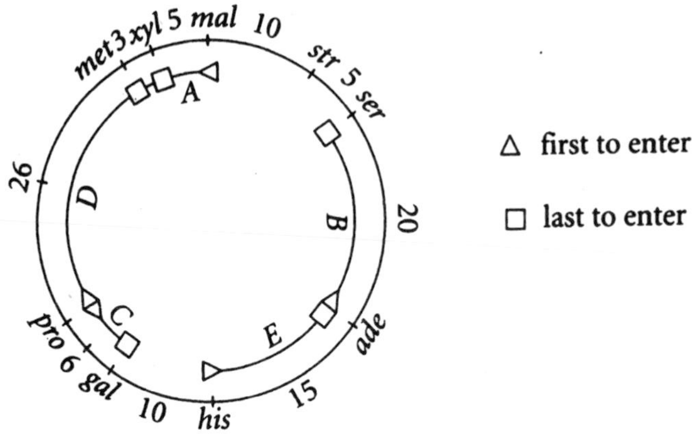
met 3 xyl 5 mal 10
str 5 ser
20
ade
15
his
10
gal
6
pro
26
A
B
C
D
E
△ first to enter
□ last to enter

c. A: Select for *mal*$^+$
B: Select for *ade*$^+$
C: Select for *pro*$^+$
D: Select for *pro*$^+$
E: Select for *his*$^+$

14. **a.** If the two genes are far enough apart to be located on separate DNA fragments, then the frequency of double transformants should be the product of the frequency of the two single transformants, or (4.3%) × (0.40%) = 0.017%. The observed double transformant frequency is 0.17 percent, a factor of 10 greater than expected. Therefore, the two genes are located close enough together to be cotransformed at a rate of 0.17 percent.

b. Here, when the two genes must be contained on separate pieces of DNA, the rate of cotransformation is much lower, confirming the conclusion in part (a).

15. The expected number of double recombinants is (0.01)(0.002)(100,000) = 2. Interference = 1 – (observed DCO/expected DCO) = 1 – 5/2 = –1.5. By definition, the interference is negative.

16. **a.** The parental genotypes are + + + and *m r tu*. For determining the *m–r* distance, the recombinant progeny are

m + tu	162
m + +	520
+ r tu	474
+ r +	172
	1328

Therefore, the map distance is 100%(1328)/10,342 = 12.8 m.u.

Using the same approach, the *r–tu* distance is 100%(2152)/10,342 = 20.8 m.u., and the *m–tu* distance is 100%(2812)/10,342 = 27.2 m.u.

b.   Because genes *m* and *tu* are the farthest apart, the gene order must be *m r tu*.

c.   The coefficient of coincidence (c.o.c.) compares the actual number of double crossovers to the expected number, (where c.o.c. = observed double crossovers/expected double crossovers). For these data, the expected number of double recombinants is (0.128)(0.208)(10,342) = 275. Thus, c.o.c. = (162 + 172)/275 = 1.2. This indicates that there are more double crossover events than predicted and suggests that the occurrence of one crossover makes a second crossover between the same DNA molecules more likely to occur.

17. a.   I: minimal plus proline and histidine
II: minimal plus purines and histidine
III: minimal plus purines and proline

b.   The order can be deduced from cotransfer rates. It is *pur–his–pro*.

c.   The closer the two genes, the higher the rate of cotransfer. *his* and *pro* are closest.

d.   *pro*$^+$ transduction requires a crossover on both sides of the *pro* gene. Because *his* is closer to *pro* than *pur*, you get the following:

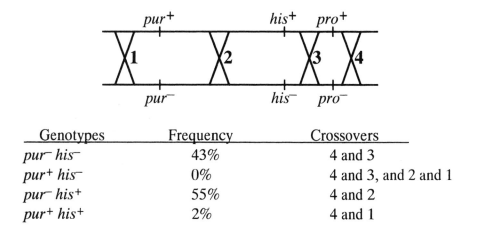

Genotypes	Frequency	Crossovers
pur⁻ his⁻	43%	4 and 3
pur⁺ his⁻	0%	4 and 3, and 2 and 1
pur⁻ his⁺	55%	4 and 2
pur⁺ his⁺	2%	4 and 1

As can be seen, a *pur+ his⁻ pro+* genotype requires four crossovers and as expected, would occur less frequently (in this example, 0%).

18. In a small percent of the cases, *gal+* transductants can arise by recombination between the *gal+* DNA of the λdgal transducing phage and the *gal⁻* gene on the chromosome. This will generate *gal+* transductants without phage integration.

19. **a.** This appears to be specialized transduction. It is characterized by the transduction of specific markers based on the position of the integration of the prophage. Only those genes near the integration site are possible candidates for misincorporation into phage particles that then deliver this DNA to recipient bacteria.

    b. The only media that supported colony growth were those lacking either cysteine or leucine. These selected for *cys+* or *leu+* transductants and indicate that the prophage is located in the *cys-leu* region.

20. **a.** This is simply calculated as the percentage of *pur+* colonies that are also *nad+*

    $$= 100\%(3 + 10)/50 = 26\%$$

    b. This is calculated as the percentage of *pur+* colonies that are also *pdx⁻*

    $$= 100\%(10 + 13)/50 = 46\%$$

    c. *pdx* is closer, as determined by cotransduction rates.

    d. From the cotransduction frequencies, you know that *pdx* is closer to *pur* than *nad* is, so there are two gene orders possible: *pur pdx nad* or *pdx pur nad*. Now, consider how a bacterial chromosome that is *pur+ pdx+ nad+* might be generated, given the two gene orders: if *pdx* is in the middle, 4 crossovers are required to get *pur+ pdx+ nad+*; if pur is in the middle, only 2 crossovers are required (see next page). The results indicate that there are fewer *pur+ pdx+ nad+* transductants than any other class suggesting that this class, is "harder" to generate than the others. This implies that *pdx* is in middle and the gene order is *pur pdx nad*.

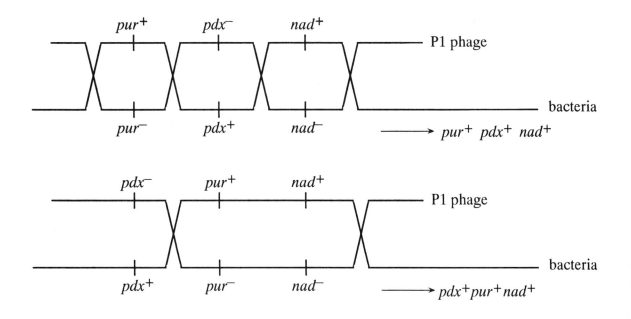

21. **a.** Owing to the medium used, all colonies are *cys⁺* but either + or − for the other two genes.

   b. (1) *cys⁺ leu⁺ thr⁺* and *cys⁺ leu⁺ thr⁻* (supplemented with threonine)
   (2) *cys⁺ leu⁺ thr⁺* and *cys⁺ leu⁻ thr⁺* (supplemented with leucine)
   (3) *cys⁺ leu⁺ thr⁺* (no supplements)

   c. Because none grew on minimal medium, no colony was *leu⁺ thr⁺*. Therefore, medium (1) had *cys⁺ leu⁺ thr⁻*, and medium (2) had *cys⁺ leu⁻ thr⁺*. The remaining cultures were *cys⁺ leu⁻ thr⁻*, and this genotype occurred in 100% − 56% − 5% = 39% of the colonies.

   d. *cys* and *leu* are cotransduced 56 percent of the time, while *cys* and *thr* are cotransduced only 5 percent of the time. This indicates that *cys* is closer to *leu* than it is to *thr*. Because no *leu⁺ cys⁺ thr⁺* cotransductants are found, it indicates that *cys* is in the middle.

22. Prototrophic strains of *E. coli* will grow on minimal media, while auxotrophic strains will only grow on media supplemented with the required molecule(s). Thus, strain 3 is prototrophic (wild-type), strain 4 is *met⁻*, strain 1 is *arg⁻*, and strain 2 is *arg⁻ met⁻*.

23. **Unpacking the Problem**

   1. *E. coli* is a bacterium and a prokaryote.

2. *E. coli* can be grown in suspension or on an agar medium. The latter method allows for the identification of individual colonies, each a clone of descendants from a single cell (and visible to the naked eye when it reaches more than 10^7 cells).

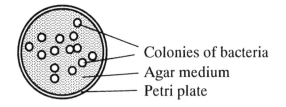

Colonies of bacteria
Agar medium
Petri plate

3. Naturally, *E. coli* is an enteric bacterium living symbiotically within the gut of host organisms (like us).

4. Minimal medium consists of inorganic salts, a carbon source for energy, and water.

5. Prototroph refers to the wild-type phenotype, or in other words, an organism that can grow on minimal media. Auxotroph refers to a mutant that can grow only on a medium supplemented with one or more specific nutrients not required by the wild-type strain.

6. In this experiment, the Hfr and the exconjugants that can grow on minimal medium are prototrophs, whereas the recipient F⁻ and the exconjugants that do not grow on minimal medium are auxotrophs.

7. Unknown strains would be grown as individual colonies on medium enriched with proline and thiamine, and then cells from each colony could be picked (by a sterile toothpick, for example) and placed individually onto medium supplemented with either thiamine or proline or onto minimal medium. Proline and thiamine auxotrophs would be identified on the basis of growth patterns. For example, a *pro⁻* strain will grow only on medium supplemented with proline.
    Instead of the labor-intensive method of individually picking cells, replica plating can be used to transfer some cells of each colony from a master plate (supplemented with proline and thiamine) to plates that contain the various media described above. The physical arrangement (and positional patterns) of colonies is used to identify the various colonies as they are transferred from plate to plate.

8. Proline is an amino acid and thiamine is a B_1 vitamin. Their chemical nature does not matter to the experiment other than that they are necessary chemicals for cell growth that prototrophs can synthesize from ingredients in minimal medium and specific auxotrophic mutants cannot.

9.

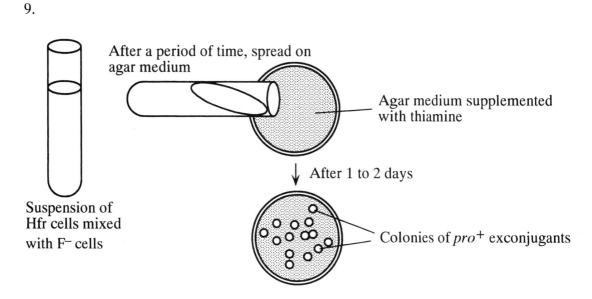

After a period of time, spread on agar medium

Agar medium supplemented with thiamine

Suspension of Hfr cells mixed with F⁻ cells

After 1 to 2 days

Colonies of *pro*⁺ exconjugants

10. Interrupted-mating experiments are used to roughly map genes onto the circular bacterial chromosome.

11. The Hfr and F⁻ strains are mixed together in solution, and then at various times, samples are removed and put into a kitchen blender, vortexed (blender is turned on) for a few seconds to disrupt conjugation, and then plated onto a medium containing the appropriate supplements. The amount of time that has passed from the mixing of the strains to mating disruption is used as a measurement for mapping. The time of first appearance of a specific gene from the Hfr in the F⁻ cell gives the gene's relative position in minutes. Typically, the F⁻ cells are streptomycin-resistant and the Hfr cells are streptomycin-sensitive. The antibiotic is used in the various media to kill the Hfr cells (which are otherwise prototrophic) and allow only those F⁻ cells that have received the appropriate gene or genes from the Hfr to grow. In this case, it would be discovered that some of the F⁻ cells would become *thi*⁺ in samples taken earlier in the experiment than samples taken when they first become *pro*⁺.

12. In this experiment, there is no attempt to disrupt conjugation. The two strains are mixed and at some later (unspecified) time, plated onto medium containing thiamine. This selects for strains that are *pro*⁺, because proline is not present in this medium.

13. Exconjugants are recipient cells (F⁻) that now contain alleles from the donor (Hfr). Typically, the F⁻ cells are streptomycin-resistant and the Hfr cells are streptomycin-sensitive. The antibiotic is used in the various media to kill the Hfr cells and allow only the appropriate F⁻ exconjugants to grow.

14. The statement "*pro* enters after *thi*" is one of gene position and order relative to the transfer of the bacterial chromosome by a particular Hfr. For the Hfr in this experiment, transfer occurs such that the *pro* gene is transferred after the *thi* gene. Because this Hfr is also *pro⁺*, it is this specific allele that is entering.

15. In this experiment, "fully supplemented" medium contains proline and thiamine.

16. All exconjugants are *pro⁺*, because that is the way they were selected. Thus, those that do not grow on minimal medium must require thiamine.

17. Genetic exchange in prokaryotes does not take place between two whole genomes as it does in eukaryotes. It takes place between one complete circular genome, the F⁻, and an incomplete linear genomic fragment donated by the Hfr. In this way, exchange of genetic information is nonreciprocal (from Hfr to F⁻). Only even numbers of crossovers are allowed between the two DNAs, because the circular chromosome would become linear otherwise. This results in unidirectional exchange, because part of the DNA of the recipient chromosome is replaced by the DNA of the donor, while the other product (the rest of the donor DNA now with some recombined recipient DNA) is nonviable and lost.

18. In this experiment, the map distance is calculated by selecting for the last marker to enter (in this case *pro⁺*) and then determining how often the earlier unselected marker (in this case *thi⁺*) is also present. Look at the following diagram:

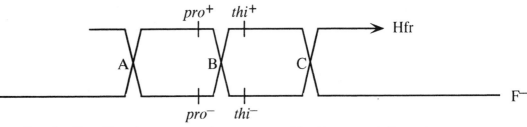

For the F⁻ cell to become *pro⁺*, two recombination events have to occur—one in the region to the left (marked A) and one in either region to the right (marked B or C.) Thus the percentage of *pro⁺* (second recombination within either B or C) that are *thi⁻* (second event only within region B) can be used to determine map distance where 1% = 1 map unit. (The map units calculated this way cannot be combined with other map unit calculations [from other experiments] to build a larger genomic map. The map units obtained are just giving an estimate of the relative sizes of intervals B and C as targets for crossover.)

Solution to the Problem

a. The two genotypes being cultured are *pro⁺ thi⁻* (grows only on media supplemented with thiamine) and *pro⁺ thi⁺* (grows on minimal media).

b. Two recombination events must occur, one on either side of *pro* (because exconjugants were plated on medium supplemented with thiamine, only *pro⁺* cells would have grown). The *pro⁺ thi⁻* strains would have had recombination in regions labeled A and B, and the *pro⁺ thi⁺* strains would have had recombination in regions labeled A and C.

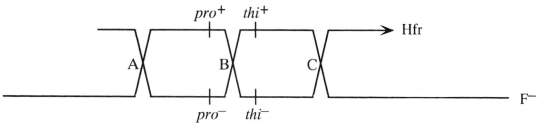

c. The distance between *pro* and *thi* is:

$$= \frac{100\%(\text{the number of colonies that are } pro^+ \ thi^-)}{\text{total number of } pro^+ \text{ colonies}}$$

$$= 100\%(40)/360 = 11.1\%$$

24. No. Closely linked loci would be expected to be cotransduced; the greater the cotransduction frequency, the closer the loci are. Because only 1 of 858 *metE⁺* was also *pyrD⁺*, the genes are not closely linked. The lone *metE⁺ pyrD⁺* could be the result of cotransduction, or it could be a spontaneous mutation of *pyrD* to *pyrD⁺*, or the result of coinfection by two separate transducing phages.

25. The *metF⁺* colonies that are now also *argC⁻* are the result of cotransduction of the two markers from the donor strain. This will be less likely than the transduction of just the *metF⁺* allele and, in these cases, the recipient remains *argC⁺*.

The following diagram illustrates the possible recombination events that will result in *metF⁺* transductants that remain *argC⁺*.

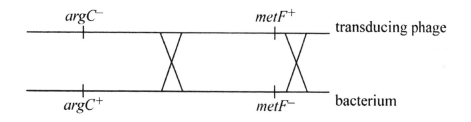

The following diagram illustrates the possible recombination events that will result in *metF*⁺ transductants that are also now *argC*⁻.

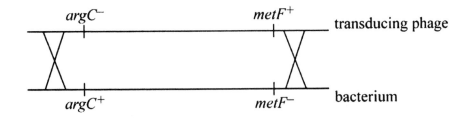

CHALLENGING PROBLEMS

26. To interpret the data, the following results are expected:

Cross	Result
F⁺ × F⁻	(L) low number of recombinants
Hfr × F⁻	(M) many recombinants
Hfr × Hfr	(0) no recombinants
Hfr × F⁺	(0) no recombinants
F⁺ × F⁺	(0) no recombinants
F⁻ × F⁻	(0) no recombinants

The only strains that show both the (L) and the (M) result when crossed are 2, 3, and 7. These must be F⁻ because that is the only cell type that can participate in a cross and give either recombination result. Hfr strains will result in only (M) or (0), and F⁺ will result in only (L) or (0) when crossed. Thus, strains 1 and 8 are F⁺, and strains 4, 5, and 6 are Hfr.

27. a.

Agar type	Selected genes
1	*c*⁺
2	*a*⁺
3	*b*⁺

b. The order of genes is revealed in the sequence of colony appearance. Because colonies first appear on agar type 1, which selects for *c*⁺, *c* must be first. Colonies next appear on agar type 3, which selects for *b*⁺, indicating that *b* follows *c*. Allele *a*⁺ appears last. The gene order is *c b a*. The three genes are roughly equally spaced.

c. In this problem you are looking at the cotransfer of the selected gene with the *d*⁻ allele (both from the Hfr). Cells that are *d*⁻ do not grow because the medium is lacking D and selecting for those cells that are *d*⁺. Therefore, the farther a gene is from gene *d*, the less likely cotransfer of the selected gene will occur with d^- and the more likely that colonies will grow (remain d^+). From the data, d is closest to b (only $^8/_{100}$ did not cotransfer d^- with b^+.) It is also closer to a than it is to c. Thus the gene order is $c\ b\ d\ a$ (or $a\ d\ b\ c$).

d. With no A or B in the agar, the medium selects for $a^+\ b^+$, and the first colonies should appear at about 17.5 minutes.

28. **a.** To survive on the selective medium, all cultures must be ery^r. Keep in mind that cells from all 300 colonies were each tested under four separate conditions.

263 colonies grew when only erythromycin is added, so these must be arg^+ $aro^+\ ery^r$. The remaining 37 cultures are mutant for one or both genes. One additional colony grew if arginine was also added to the medium ($264 - 263 = 1$). It must be $arg^-\ aro^+\ ery^r$. A total of 290 colonies are arg^+ because they grew when erythromycin and aromatic amino acids were added to the medium. Of these, 27 are aro^- ($290 - 263 = 27$). The genotypes and their frequencies are summarized below:

$$
\begin{array}{rl}
263 & ery^r\ arg^+\ aro^+ \\
27 & ery^r\ arg^+\ aro^- \\
1 & ery^r\ arg^-\ aro^+ \\
\underline{9} & ery^r\ arg^-\ aro^- \\
300 &
\end{array}
$$

b. Recombination in the *aro–arg* region is represented by two genotypes: aro^+ arg^- and $aro^-\ arg+$. The frequency of recombination is

$$100\%(1 + 27)/300 = 9.3 \text{ m.u.}$$

Recombination in the ery–arg region is represented by two genotypes: aro^+ arg^- and $aro^-\ arg^-$. The frequency of recombination is

$$100\%(1 + 9)/300 = 3.3 \text{ m.u.}$$

Recombination in the *ery–aro* region is represented by three genotypes: $arg^+\ aro^-$, $arg^-\ aro^-$, and $arg^-\ aro^+$. Recall that the DCO must be counted twice. The frequency of recombination is

$$100\%(27 + 9 + 2)/300 = 12.6 \text{ m.u.}$$

c. The ratio is 28:10, or 2.8:1.0.

29. **a.** To determine which genes are close, compare the frequency of double transformants. Pairwise testing gives low values whenever B is involved but fairly high rates when any drug but B is involved. This suggests that the gene for B resistance is not close to the other three genes.

b. To determine the relative order of genes for resistance to A, C, and D, compare the frequency of double and triple transformants. The frequency of resistance to AC is approximately the same as resistance to ACD. This strongly suggests that D is in the middle. Also, the frequency of AD co-resistance is higher than AC (suggesting that the gene for A resistance is closer to D than to C), and the frequency of CD is higher than AC (suggesting that C is closer to D than to A).

30. **a. and b.**

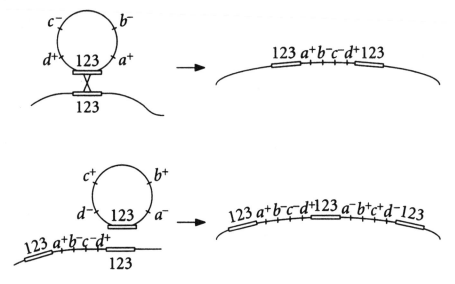

c.

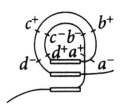

d.

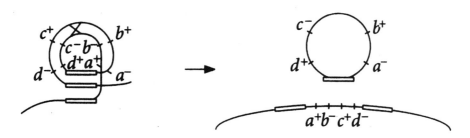

31. If *trp1* and *trp2* are alleles of the same locus, then a cross between strains A and B will not result in Trp⁺ cells; if they are not allelic, strain B cells that have received the F′ from strain A will be Trp+ (complementation).

32. If a compound is not added and growth occurs, the *E. coli* recipient cell must have received the wild-type genes for production of those nutrients by transduction. Thus, the BCE culture selects for cells that are now a^+ and d^+, the BCD culture selects for cells that are a^+ and e^+, and the ABD culture selects for cells that are c^+ and e^+. These genes can be aligned (see below) to give the map order of *d a e c*. (Notice that *b* is never cotransduced and is therefore distant from this group of genes.)

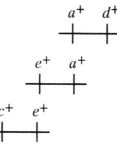

33. To isolate the specialized transducing particles of phage φ80 that carried *lac⁺*, the researchers would have had to lysogenize the strain with φ80, induce the phage with UV, and then use these lysates to transduce a Lac⁻ strain to Lac⁺. Lac⁺ colonies would then be used to make a new lysate, which should be highly enriched for the *lac⁺* transducing phage.

34. *atoE* may encode a protein necessary for the transport of the food dye into the cell. All cells are surrounded by a semipermeable membrane so many substances require specific transport proteins to get into or out of the cell. Other possible white colonies expected are those in which other genes impacting uptake have been mutated or spontaneous mutants of the *atoE* gene.

6

Gene Interaction

BASIC PROBLEMS

1. Lactose is composed of one molecule of galactose and one molecule of glucose. A secondary cure would result if all galactose and lactose were removed from the diet. The disorder would be expected not to be dominant, because one good copy of the gene should allow for at least some, if not all, breakdown of galactose. In fact, the disorder is recessive.

2. Assuming homozygosity for the normal gene, the mating is $A/A \cdot b/b \times a/a \cdot B/B$. The children would be normal, $A/a \cdot B/b$ (see Problem 12).

3.

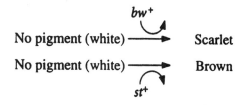

Scarlet plus brown results in red.

4. Growth will be supported by a particular compound if it is later in the pathway than the enzymatic step blocked in the mutant. Restated, the more mutants a compound supports, the later in the pathway it must be. In this example, compound G supports growth of all mutants and can be considered the end product of the pathway. Alternatively, compound E does not support the growth of any mutant and can be considered the starting substrate for the pathway. The data indicate the following:

    a. and b.

    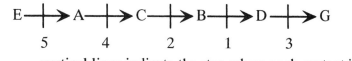

    vertical lines indicate the step where each mutant is blocked

c. A heterokaryon of double mutants 1, 3 and 2, 4 would grow as the first would supply functional 2 and 4, and the second would supply functional 1 and 3.

A heterokaryon of the double mutants 1, 3 and 3, 4 would not grow as both are mutant for 3.

A heterokaryon of the double mutants 1, 2 and 2, 4 and 1, 4 would grow as the first would supply functional 4, the second would supply functional 1, and the last would supply functional 2.

5. a. If enzyme A was defective or missing (m_2/m_2), red pigment would still be made and the petals would be red.

b. Purple, because it has a wild-type allele for each gene, and you are told that the mutations are recessive.

c.

9	$M_1/-$; $M_2/-$	purple
3	m_1/m_1 ; $M_2/-$	blue
3	$M_1/-$; m_2/m_2	red
1	m_1/m_1 ; m_2/m_2	white

d. The mutant alleles do not produce functional enzyme. However, enough functional enzyme must be produced by the single wild-type allele of each gene to synthesize normal levels of pigment.

6. a. If enzyme B is missing, a white intermediate will accumulate and the petals will be white.

b. If enzyme D is missing, a blue intermediate will accumulate and the petals will be blue.

c.

P b/b ; D/D × B/B ; d/d

F_1 B/b ; D/d purple

d.

P b/b ; D/D × B/B ; d/d

F_1 B/b ; D/d × B/b ; D/d

F_2 9 $B/-$; $D/-$ purple
3 b/b ; $D/-$ white
3 $B/-$; d/d blue
1b/b ; d/d white

The ratio of purple : blue : white would be 9:3:4.

7. The woman must be A/O, so the mating is $A/O \times A/B$. Their children will be

Genotype	Phenotype
$1/4\ A/A$	A
$1/4\ A/B$	AB
$1/4\ A/O$	A
$1/4\ B/O$	B

8. You are told that the cross of two erminette fowls results in 22 erminette, 14 black, and 12 pure white. Two facts are important: (1) the parents consist of only one phenotype, yet the offspring have three phenotypes, and (2) the progeny appear in an approximate ratio of 1:2:1. These facts should tell you immediately that you are dealing with a heterozygous × heterozygous cross involving one gene and that the erminette phenotype must be the heterozygous phenotype.

When the heterozygote shows a different phenotype from either of the two homozygotes, the heterozygous phenotype results from incomplete dominance or codominance. Because two of the three phenotypes contain black, either fully or in an occasional feather, you might classify erminette as an instance of incomplete dominance because it is intermediate between fully black and fully white. Alternatively, because erminette has both black and white feathers, you might classify the phenotype as codominant. Your decision will rest on whether you look at the whole animal (incomplete dominance) or at individual feathers (codominance). This is yet another instance where what you conclude is determined by how you observe.

To test the hypothesis that the erminette phenotype is a heterozygous phenotype, you could cross an erminette with either, or both, of the homozygotes. You should observe a 1:1 ratio in the progeny of both crosses.

9. a. The original cross and results were

P   long, white × round, red

F_1 oval, purple

F_2	9 long, red	19 oval, red	8 round, white
	15 long, purple	32 oval, purple	16 round, purple
	8 long, white	16 oval, white	9 round, red
	32 long	67 oval	33 round

The data show that, when the results are rearranged by shape, a 1:2:1 ratio is observed for color within each shape category. Likewise, when the data are rearranged by color, a 1:2:1 ratio is observed for shape within each color category.

9 long, red	15 long, purple	8 round, white
19 oval, red	32 oval, purple	16 oval, white
9 round, red	16 round, purple	8 long, white
37 red	63 purple	32 white

A 1:2:1 ratio is observed when there is a heterozygous × heterozygous cross. Therefore, the original cross was a dihybrid cross. Both oval and purple must represent an incomplete dominant phenotype.

Let L = long, L' = round, R = red and R' = white. The cross becomes

P $\quad L/L\, ;\, R'/R' \times L'/L'\, ;\, R/R$

$F_1 \quad L/L'\, ;\, R/R' \times L/L'\, ;\, R/R'$

F_2

	$^1/_4\, L/L \times$	$^1/_4\, R/R =$	$^1/_{16}$	long, red
		$^1/_2\, R/R' =$	$^1/_8$	long, purple
		$^1/_4\, R'/R' =$	$^1/_{16}$	long, white
	$^1/_2\, L/L' \times$	$^1/_4\, R/R =$	$^1/_8$	oval, red
		$^1/_2\, R/R' =$	$^1/_4$	oval, purple
		$^1/_4\, R'/R' =$	$^1/_8$	oval, white
	$^1/_4\, L'/L' \times$	$^1/_4\, R/R =$	$^1/_{16}$	round, red
		$^1/_2\, R/R' =$	$^1/_8$	round, purple
		$^1/_4\, R'/R' =$	$^1/_{16}$	round, white

b. A long, purple × oval, purple cross is as follows

P $\quad L/L\, ;\, R/R' \times L/L'\, ;\, R/R''$

F_1

	$^1/_2\, L/L \times$	$^1/_4\, R/R =$	$^1/_8$ long, red
		$^1/_2\, R/R' =$	$^1/_4$ long, purple
		$^1/_4\, R'/R' =$	$^1/_8$ long, white
	$^1/_2\, L/L' \times$	$^1/_4\, R/R =$	$^1/_8$ oval, red
		$^1/_2\, R/R' =$	$^1/_4$ oval, purple
		$^1/_4\, R'/R' =$	$^1/_8$ oval, white

10. From the cross $c^+/c^{ch} \times c^{ch}/c^{ch}$ the progeny are

$^1/_4 \quad c^+/c^{ch} \qquad$ full color

$1/4$ c^+/c^h full color
$1/4$ c^{ch}/c^{ch} chinchilla
$1/4$ c^{ch}/c^h chinchilla

Thus, 50 percent of the progeny will be chinchilla.

11.   **a.**   The data indicate that there is a single gene with multiple alleles. All the ratios produced are 3:1 (complete dominance), 1:2:1 (incomplete of codominance), or 1:1 (test cross). The order of dominance is

$$\text{black } (C^b) > \text{sepia } (C^s) > \text{cream } (C^c) > \text{albino } (C^a)$$

Cross	Parents	Progeny	Conclusion
Cross 1:	$C^b/C^a \times C^b/C^a$	$3\ C^b/-\ :\ 1\ C^a/C^a$	C^b is dominant to C^a.
Cross 2:	$C^b/C^s \times C^a/C^a$	$1\ C^b/C^a :\ 1\ C^s/C^a$	C^b is dominant to C^s; C^s is dominant to C^a.
Cross 3:	$C^c/C^a \times C^c/C^a$	$3\ C^c/-\ :\ 1\ C^a/C^a$	C^c is dominant to C^a.
Cross 4:	$C^s/C^a \times C^c/C^a$	$1\ C^c/C^a :\ 2\ C^s/-\ :\ 1\ C^a/C^a$	C^s is dominant to C^c.
Cross 5:	$C^b/C^c \times C^a/C^a$	$1\ C^b/C^a :\ 1\ C^c/C^a$	C^b is dominant to C^c.
Cross 6:	$C^b/C^s \times C^c/-$	$1\ C^b/-\ :\ 1\ C^s/-$	"–" can be C^c or C^a.
Cross 7:	$C^b/C^s \times C^s/-$	$1\ C^b/C^s :\ 1\ C^s/-$	"–" can be C^c or C^a.
or	$C^b/- \times C^s/C^s$	$1\ C^b/C^s :\ 1\ C^s/-$	"–" can be C^s, C^c or C^a.
Cross 8:	$C^b/C^c \times C^s/C^c$	$1\ C^s/C^c :\ 2\ C^b/-\ :\ 1\ C^c/C^c$	
Cross 9:	$C^s/C^c \times C^s/C^c$	$3\ C^s/-\ :\ 1\ C^c/C^c$	
Cross 10:	$C^c/C^a \times C^a/C^a$	$1\ C^c/C^a :\ 1\ C^a/C^a$	

   b.   The progeny of the cross $C^b/C^s \times C^b/C^c$ will be $3/4$ black ($1/4$ C^b/C^b, $1/4$ C^b/C^c, $1/4$ C^b/C^s) : $1/4$ sepia (C^s/C^c). The progeny of the cross C^b/C^a (or C^b/C^c) $\times C^b/C^c$ will be $3/4$ black: $1/4$ cream.

12.   Both codominance (=) and classical dominance (>) are present in the multiple allelic series for blood type: $A = B$, $A > O$, $B > O$.

Parents' phenotype	Parents' possible genotypes	Parents' possible children
a. AB × O	$A/B \times O/O$	$A/O, B/O$
b. A × O	A/A or $A/O \times O/O$	$A/O, O/O$
c. A × AB	A/A or $A/O \times A/B$	$A/A, A/B, A/O, B/O$
d. O × O	$O/O \times O/O$	O/O

The possible genotypes of the children are

Phenotype	Possible genotypes
O	*O/O*
A	*A/A, A/O*
B	*B/B, B/O*
AB	*A/B*

Using the assumption that each set of parents had one child, the following combinations are the only ones that will work as a solution.

Parents	Child
a. AB × O	B
b. A × O	A
c. A × AB	AB
d. O × O	O

13. *M* and *N* are codominant alleles. The rhesus group is determined by classically dominant alleles. The *ABO* alleles are mixed codominance and classical dominance (see Problem 6).

Person	Blood type			Possible paternal contribution		
husband	O	M	Rh⁺	*O*	*M*	*R* or *r*
wife's lover	AB	MN	Rh⁻	*A* or *B*	*M* or *N*	*r*
wife	A	N	Rh⁺	*A* or *O*	*N*	*R* or *r*
child 1	O	MN	Rh⁺	*O*	*M*	*R* or *r*
child 2	A	N	Rh⁺	*A* or *O*	*N*	*R* or *r*
child 3	A	MN	Rh⁻	*A* or *O*	*M*	*r*

The wife must be *A/O* ; *N/N* ; *R/r*. (She has a child with type O blood and another child that is Rh⁻ so she must carry both of these recessive alleles.) Only the husband could donate O to child 1. Only the lover could donate A and N to child 2. Both the husband and the lover could have donated the necessary alleles to child 3.

14. The key to solving this problem is in the statement that breeders cannot develop a pure–breeding stock and that a cross of two platinum foxes results in some normal progeny. Platinum must be dominant to normal color and heterozygous (*A/a*). An 82:38 ratio is very close to 2:1. Because a 1:2:1 ratio is expected in a heterozygous cross, one genotype is nonviable. It must be the *A/A*, homozygous platinum, genotype that is nonviable, because the homozygous recessive genotype is normal color (*a/a*). Therefore, the platinum allele is a pleiotropic allele that governs coat color in the heterozygous state and is lethal when homozygous.

15. **a.** Because Pelger crossed with normal stock results in two phenotypes in a 1:1 ratio, either Pelger or normal is heterozygous (A/a) and the other is homozygous (a/a) recessive. The problem states that normal is true–breeding, or a/a. Pelger must be A/a.

b. The cross of two Pelger rabbits results in three phenotypes. This means that the Pelger anomaly is dominant to normal. This cross is $A/a \times A/a$, with an expected ratio of 1:2:1. Because the normal must be a/a, the extremely abnormal progeny must be A/A. There were only 39 extremely abnormal progeny because the others died before birth.

c. The Pelger allele is pleiotropic. In the heterozygous state, it is dominant for nuclear segmentation of white blood cells. In the homozygous state, it is a lethal.

You could look for the nonsurviving fetuses in utero. Because the hypothesis of embryonic death when the Pelger allele is homozygous predicts a one-fourth reduction in litter size, you could also do an extensive statistical analysis of litter size, comparing normal × normal with Pelger × Pelger.

d. By analogy with rabbits, the absence of a homozygous Pelger anomaly in humans can be explained as recessive lethality. Also, because one in 1000 people have the Pelger anomaly, a heterozygous × heterozygous mating would be expected in only one of 1 million.

($^1/_{1000} \times {}^1/_{1000}$) random matings, and then only one in four of the progeny would be expected to be homozygous. Thus, the homozygous Pelger anomaly is expected in only 1 of 4 million births. This is extremely rare and might not be recognized.

e. By analogy with rabbits, among the children of a man and a woman with the Pelger anomaly, two-thirds of the surviving progeny would be expected to show the Pelger anomaly and one-third would be expected to be normal. The developing fetus that is homozygous for the Pelger allele would not be expected to survive until birth.

16. **a.** The sex ratio is expected to be 1:1.

b. The female parent was heterozygous for an X-linked recessive lethal allele. This would result in 50 percent fewer males than females.

c. Half of the female progeny should be heterozygous for the lethal allele and half should be homozygous for the nonlethal allele. Individually mate the F_1 females and determine the sex ratio of their progeny.

17. Note that a cross of the short-bristled female with a normal male results in two phenotypes with regard to bristles and an abnormal sex ratio of two females : one male. Furthermore, all the males are normal, while the females are normal and short in equal numbers. Whenever the sexes differ with respect to phenotype among the progeny, an X-linked gene is implicated. Because only the normal phenotype is observed in males, the short-bristled phenotype must be heterozygous, and the allele must be a recessive lethal. Thus the first cross was $A/a \times a/Y$.

Long-bristled females (a/a) were crossed with long-bristled males (a/Y). All their progeny would be expected to be long-bristled (a/a or a/Y). Short-bristled females (A/a) were crossed with long-bristled males (a/Y). The progeny expected are

$1/4$ A/a	short-bristled females
$1/4$ a/a	long-bristled females
$1/4$ a/Y	long-bristled males
$1/4$ A/Y	nonviable

18. In order to do this problem, you should first restate the information provided. The following two genes are independently assorting

h/h = hairy	s/s = no effect
H/h = hairless	S/s suppresses H/h, giving hairy
H/H = lethal	S/S = lethal

a. The cross is H/h ; $S/s \times H/h$; S/s. Because this is a typical dihybrid cross, the expected ratio is 9:3:3:1. However, the problem cannot be worked in this simple fashion because of the epistatic relationship of these two genes. Therefore, the following approach should be used.

For the H gene, you expect $1/4$ H/H : $1/2$ H/h : $1/4$ h/h. For the S gene, you expect $1/4$ S/S : $1/2$ S/s : $1/4$ s/s. To get the final ratios, multiply the frequency of the first genotype by the frequency of the second genotype.

$1/4$ H/H all progeny die regardless of the S gene

		$1/4$ S/S =	$1/8$ H/h ; S/S	die
$1/2$ H/h	×	$1/2$ S/s =	$1/4$ H/h ; S/s	hairy
		$1/4$ s/s =	$1/8$ H/h ; s/s	hairless
		$1/4$ S/S =	$1/16$ h/h ; S/S	die
$1/4$ h/h	×	$1/2$ S/s =	$1/8$ h/h ; S/s	hairy
		$1/4$ s/s =	$1/16$ h/h ; s/s	hairy

Of the $^9/_{16}$ living progeny, the ratio of hairy to hairless is 7:2.

b. This cross is H/h ; $s/s \times H/h$; S/s. A 1:2:1 ratio is expected for the H gene and a 1:1 ratio is expected for the S gene.

$^1/_4$ H/H			all progeny die regardless of the S gene	

$^1/_2$ H/h $\times$ $^1/_2$ S/s = $^1/_4$ H/h ; S/s hairy
$^1/_2$ s/s = $^1/_4$ H/h ; s/s hairless

$^1/_4$ h/h $\times$ $^1/_2$ S/s = $^1/_8$ h/h ; S/s hairy
$^1/_2$ s/s = $^1/_8$ h/h ; s/s hairy

Of the $^3/_4$ living progeny, the ratio of hairy to hairless is 2:1.

19. a. The mutations are in two different genes as the heterokaryon is prototrophic (the two mutations complemented each other).

b. $leu1^+$; $leu\ 2^-$ and $leu1^-$; $leu2^+$

c. With independent assortment, expect

$^1/_4$ $leu1^+$; $leu\ 2^-$
$^1/_4$ $leu1^-$; $leu2^+$
$^1/_4$ $leu1^-$; $leu\ 2^-$
$^1/_4$ $leu1^+$; $leu\ 2^+$

20. a. The first type of prototroph is due to reversion of the original $ade1$ mutant to wild type. The second type of prototroph is due to a new mutation (call it sup^m) in an unlinked gene that suppresses the $ade1^-$ phenotype. For the results of crossing type 2 prototrophs to wild type

$ade1^-$; $sup^m \times ade1^+$; sup

and the results would be

$^1/_4$ $ade1^+$; sup prototroph
$^1/_4$ $ade1^+$; sup^m prototroph
$^1/_4$ $ade1^-$; sup^m prototroph
$^1/_4$ $ade1^-$; sup auxotroph

b. type 1 would be $ade1^+$; sup
type 2 would be $ade1^-$; sup^m

c.   For the results of crossing type 2 prototrophs to the original mutant

$$adel^- \; ; \; sup^m \; \times \; adel^- \; ; \; sup$$

and the results would be

$1/2$ $adel^-$; sup auxotroph
$1/2$ $adel^-$; sup^m prototroph

21. **a.** colorless

b. magenta

c. colorless

d. p/p ; Q/Q
P/P ; q/q
p/p ; q/q

e. 9 red : 4 colorless : 3 magenta

22. The suggestion from the data is that the two albino lines had mutations in two different genes. When the extracts from the two lines were placed in the same test tube, they were capable of producing color because the gene product of one line was capable of compensating for the absence of a gene product from the second line.

a. Grind a sample of each specimen separately (negative control) to ensure that the grinding process does not activate or release an enzyme from some compartment that allows the color change. Another control is to cross the two pure-breeding lines. The cross would be A/A ; $b/b \times a/a$; B/B. The progeny will be A/a ; B/b, and all should be reddish purple.

b. The most likely explanation is that the red pigment is produced by the action of at least two different gene products. When petals of the two plants were ground together, the different defective enzyme of each plant was complemented by the normal enzyme of the other.

c. The genotypes of the two lines would be A/A ; b/b and a/a ; B/B.

d. The F_1 would be all be pigmented, A/a ; B/b. This is an example of complementation. The mutants are defective for different genes. The F_2 would be

9    $A/-$; $B/-$ pigmented
3   a/a ; $B/-$ white

3    $A/-$; b/b white
1   a/a ; b/b white

23. a. This is an example where one phenotype in the parents gives rise to three phenotypes in the offspring. The "frizzle" is the heterozygous phenotype and shows incomplete dominance.

P A/a (frizzle) × A/a (frizzle)

F_1 1 A/A (normal) : 2 A/a (frizzle) : 1 a/a (woolly)

b. If A/A (normal) is crossed to a/a (woolly), all offspring will be A/a (frizzle).

24. *Unpacking the Problem*

1. The character being studied is petal color.

2. The wild-type phenotype is blue.

3. A variant is a phenotypic difference from wild type that is observed.

4. There are two variants: pink and white.

5. *In nature* means that the variants did not appear in laboratory stock and, instead, were found growing wild.

6. Possibly, the variants appeared as a small patch or even a single plant within a larger patch of wild type.

7. Seeds would be grown to check the outcome from each cross.

8. Given that no sex linkage appears to exist (sex is not specified in parents or offspring), *blue × white* means the same as *white × blue*. Similar results would be expected because the trait being studied appears to be autosomal.

9. The first two crosses show a 3:1 ratio in the F_2, suggesting the segregation of one gene. The third cross has a 9:4:3 ratio for the F_2, suggesting that two genes are segregating.

10. Blue is dominant to both white and pink.

11. *Complementation* refers to generation of wild-type progeny from the cross of two strains that are mutant in different genes.

12. The ability to make blue pigment requires two enzymes that are individually defective in the pink or white strains. The F_1 progeny of this cross is blue, since each has inherited one nonmutant allele for both genes and can therefore produce both functional enzymes.

13. Blueness from a pink × white cross arises through complementation.

14. The following ratios are observed: 3:1, 9:4:3.

15. There are monohybrid ratios observed in the first two crosses.

16. There is a modified 9:3:3:1 ratio in the third cross.

17. A monohybrid ratio indicates that one gene is segregating, while a dihybrid ratio indicates that two genes are segregating.

18. 15:1, 12:3:1, 9:6:1, 9:4:3, 9:7, 1:2:1, 2:1

19. There is a modified dihybrid ratio in the third cross.

20. A modified dihybrid ratio most frequently indicates the interaction of two or more genes.

21. Recessive epistasis is indicated by the modified dihybrid ratio.

22.

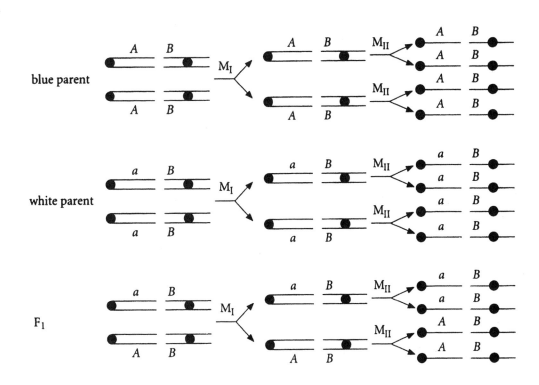

blue parent

white parent

F_1

Solution to the Problem

a. Let A = wild-type, a = white, B = wild-type, and b pink.

Cross 1:	P	blue × white	$A/A \; ; B/B \times a/a \; ; B/B$
	F_1	all blue	all $A/a \; ; B/B$
	F_2	3 blue : 1 white	$3 \, A/- \; ; B/B : 1 \, a/a \; ; B/B$
Cross 2:	P	blue × pink	$A/A \; ; B/B \times A/A \; ; b/b$
	F_1	all blue	all $A/A \; ; B/b$
	F_2	3 blue : 1 pink	$3 \, A/A \; ; B/- : 1 \, A/A \; ; b/b$
Cross 3:	P	pink × white	$A/A \; ; b/b \times a/a \; ; B/B$
	F_1	all blue	all $A/a \; ; B/b$
	F_2	9 blue	$9 \, A/- \; ; B/-$
		4 white	$3 \, a/a \; ; B/- : 1 \, a/a \; ; b/b$
		3 pink	$3 \, A/- \; ; b/b$

When the allele a is homozygous, the expression of alleles B or b is blocked or masked. The white phenotype is epistatic to the pigmented phenotypes. It is likely that the product of the A gene produces an intermediate that is then modified by the product of the B gene. If the plant is a/a, this intermediate is not made and the phenotype of the plant is the same regardless of the ability to produce functional B product.

b. The cross is

$$F_2 \quad \text{blue} \times \text{white}$$
$$F_3 \quad {}^3/_8 \text{ blue}$$
$$\quad\quad {}^1/_8 \text{ pink}$$
$$\quad\quad {}^4/_8 \text{ white}$$

Begin by writing as much of each genotype as can be assumed:

$$F_2 \quad A/- \,; B/- \;\times\; a/a \,; -/-$$
$$F_3 \quad {}^3/_8 \; A/- \,; B/-$$
$$\quad\quad {}^1/_8 \; A/- \,; b/b$$
$$\quad\quad {}^4/_8 \; a/a \,; -/-$$

Notice that both a/a and b/b appear in the F_3 progeny. In order for these homozygous recessives to occur, each parent must have at least one a and one b. Using this information, the cross becomes

$$F_2 \quad A/a \,; B/b \;\times\; a/a \,; -/b$$
$$F_3 \quad {}^3/_8 \; A/a \,; B/b$$
$$\quad\quad {}^1/_8 \; A/a \,; b/b$$
$$\quad\quad {}^4/_8 \; a/a \,; b/-$$

The only remaining question is whether the white parent was homozygous recessive, b/b, or heterozygous, B/b. If the white parent had been homozygous recessive, then the cross would have been a testcross of the blue parent, and the progeny ratio would have been

1 blue : 1 pink : 2 white, or 1 $A/a \,; B/b$: 1 $A/a \,; b/b$: 1 $a/a \,; B/b$: 1 $a/a \,; b/b$.

This was not observed. Therefore, the white parent had to have been heterozygous, and the F_2 cross was $A/a \,; B/b \times a/a \,; B/b$.

25. It is possible to produce black offspring from two pure-breeding recessive albino parents if albinism results from mutations in two different genes. If the cross is designated

$$A/A \,; b/b \times a/a \,; B/B$$

all offspring would be

$$A/a \,; B/b$$

and they would have a black phenotype because of complementation.

26. The data indicate that white is dominant to solid purple. Note that the F_2 are in an approximate 12:3:1 ratio. In order to achieve such a ratio, epistasis must be involved.

a. Because a modified 9:3:3:1 ratio was obtained in the F_2, the F_1 had to be a double heterozygote. Solid purple occurred at one-third the rate of white, which means that it will be in the form of either $D/-$; e/e or d/d ; $E/-$. In order to achieve a double heterozygote in the F_1, the original white parent also has to be either $D/-$; e/e or d/d ; $E/-$.

Arbitrarily assume that the original cross was D/D ; e/e (white) × d/d ; E/E (purple). The F_1 would all be D/d ; E/e. The F_2 would be

9 $D/-$; $E/-$	white, by definition
3 d/d ; $E/-$	purple, by definition
3 $D/-$; e/e	white, by deduction
1 d/d ; e/e	spotted purple, by deduction

Under these assumptions, D blocks the expression of both E and e. The d allele has no effect on the expression of E and e. E results in solid purple, while e results in spotted purple. It would also be correct, of course, to assume the opposite set of epistatic relationships (E blocks the expression of D or d, D results in solid purple, and d results in spotted purple).

b. The cross is white × solid purple. While the solid purple genotype must be d/d ; $E/-$, as defined in part (a), the white genotype can be one of several possibilities. Note that the progeny phenotypes are in a 1:2:1 ratio and that one of the phenotypes, spotted, must be d/d ; e/e. In order to achieve such an outcome, the purple genotype must be d/d ; E/e. The white genotype of the parent must contain both a D and a d allele in order to produce both white ($D/-$) and spotted plants (d/d). At this point, the cross has been deduced to be D/d ; $-/-$ (white) × d/d ; E/e (purple).

If the white plant is E/E, the progeny will be

$1/2$ D/d ; $E/-$	white
$1/2$ d/d ; $E/-$	solid purple

This was not observed. If the white plant is E/e, the progeny will be

$3/8$ D/d ; $E/-$	white
$1/8$ D/d ; e/e	white
$3/8$ d/d ; $E/-$	solid purple
$1/8$ d/d ; e/e	spotted purple

The phenotypes were observed, but in a different ratio. If the white plant is e/e, the progeny will be

$1/4$ D/d ; E/e white
$1/4$ D/d ; e/e white
$1/4$ d/d ; E/e solid purple
$1/4$ d/d ; e/e spotted purple

This was observed in the progeny. Therefore, the parents were D/d ; e/e (white) × d/d ; E/e (purple).

27.   **a. and b.** Crosses 1–3 show a 3:1 ratio, indicating that brown, black, and yellow are all alleles of one gene. Crosses 4–6 show a modified 9:3:3:1 ratio, indicating that at least two genes are involved. Those crosses also indicate that the presence of color is dominant to its absence. Furthermore, epistasis must be involved for there to be a modified 9:3:3:1 ratio.

By looking at the F_1 of crosses 1–3, the following allelic dominance relationships can be determined: black > brown > yellow. Arbitrarily assign the following genotypes for homozygotes: B^l/B^l = black, B^r/B^r = brown, B^y/B^y = yellow.

By looking at the F_2 of crosses 4–6, a white phenotype is composed of two categories: the double homozygote and one class of the mixed homozygote/heterozygote. Let lack of color be caused by c/c. Color will therefore be $C/-$.

Parents	F_1	F_2
1 B^r/B^r ; C/C × B^y/B^y ; C/C	B^r/B^y ; C/C	3 $B^r/-$; C/C : 1 B^y/B^y ; C/C
2 B^l/B^l ; C/C × B^r/B^r ; C/C	B^l/B^r ; C/C	3 $B^l/-$; C/C : 1 B^r/B^r ; C/C
3 B^l/B^l ; C/C × B^y/B^y ; C/C	B^l/B^y ; C/C	3 $B^l/-$; C/C : 1 B^y/B^y ; C/C
4 B^l/B^l ; c/c × B^y/B^y ; C/C	B^l/B^y ; C/c	9 $B^l/-$; $C/-$: 3 B^y/B^y ; $C/-$: 3 $B^l/-$; c/c : 1 B^y/B^y ; c/c
5 B^l/B^l ; c/c × B^r/B^r ; C/C	B^l/B^r ; C/c	9 $B^l/-$; $C/-$: 3 B^r/B^r ; $C/-$: 3 $B^l/-$; c/c : 1 B^r/B^r ; c/c
6 B^l/B^l ; C/C × B^y/B^y ; c/c	B^l/B^y ; C/c	9 $B^l/-$; $C/-$: 3 B^y/B^y ; $C/-$: 3 $B^l/-$; c/c : 1 B^y/B^y ; c/c

28. It is possible to produce normally pigmented offspring from albino parents if albinism results from mutations in either of two different genes. If the cross is designated

$$A/A \cdot b/b \times a/a \cdot B/B$$

then all the offspring would be

$$A/a \cdot B/b$$

and they would have a pigmented phenotype because of complementation.

29. The first step in each cross is to write as much of the genotype as possible from the phenotype.

Cross 1: $A/- \; ; \; B/- \times a/a \; ; \; b/b \quad \rightarrow \quad 1\,A/- \; ; \; B/- : 2\,?/a; \; ?/b : 1\,a/a \; ; \; b/b$

Because the double recessive appears, the blue parent must be $A/a \; ; \; B/b$. The $1/2$ purple then must be $A/a \; ; \; b/b$ and $a/a \; ; \; B/b$.

Cross 2: $?/? \; ; \; ?/? \times ?/? \; ; \; ?/? \quad \rightarrow \quad 1\,A/- \; ; \; B/- : 2\,?/? \; ; \; ?/? : 1\,a/a \; ; \; b/b$

The two parents must be, in either order, $A/a \; ; \; b/b$ and $a/a \; ; \; B/b$. The two purple progeny must be the same. The blue progeny are $A/a \; ; \; B/b$.

Cross 3: $A/- \; ; \; B/- \times A/- \; ; \; B/- \rightarrow \quad 3\,A/- \; ; \; B/- : 1\,?/? \; ; \; ?/?$

The only conclusions possible here are that one parent is either A/A or B/B and the other parent is B/b if the first is A/A or A/a if the first is B/B.

Cross 4: $A/- \; ; \; B/- \times ?/? \; ; \; ?/? \quad \rightarrow \quad 3\,A/- \; ; \; B/- : 4\,?/? \; ; \; ?/? : 1\,a/a \; ; \; b/b$

The purple parent can be either $A/a \; ; \; b/b$ or $a/a \; ; \; B/b$ for this answer. Assume the purple parent is $A/a \; ; \; b/b$. The blue parent must be $A/a \; ; \; B/b$. The progeny are

$$3/4\,A/- \quad \times \quad \begin{aligned} 1/2\,B/b &= 3/8\,A/- \; ; \; B/b \quad \text{blue} \\ 1/2\,b/b &= 3/8\,A/- \; ; \; b/b \quad \text{purple} \end{aligned}$$

$$1/4\,a/a \quad \times \quad \begin{aligned} 1/2\,B/b &= 1/8\,a/a \; ; \; B/b \quad \text{purple} \\ 1/2\,b/b &= 1/8\,a/a \; ; \; b/b \quad \text{scarlet} \end{aligned}$$

Cross 5: $A/- \; ; \; b/b \times a/a \; ; \; b/b \quad \rightarrow \quad 1\,A/- \; ; \; b/b : 1\,a/a \; ; \; b/b$

As written this is a testcross for gene A. The purple parent and progeny are $A/a \; ; \; b/b$. Alternatively, the purple parent and progeny could be $a/a \; ; \; B/b$.

30. The F_1 progeny of cross 1 indicate that sun red is dominant to pink. The F_2 progeny, which are approximately in a 3:1 ratio, support this. The same pattern is seen in crosses 2 and 3, with sun red dominant to orange and orange dominant to pink. Thus, we have a multiple allelic series with sun red > orange > pink. In all three crosses, the parents must be homozygous.

If c^{sr} = sun red, c^o = orange, and c^p = pink, then the crosses and the results are

Cross	Parents	F_1	F_2
1	$c^{sr}/c^{sr} \times c^p/c^p$	c^{sr}/c^p	3 $c^{sr}/-$: 1 c^p/c^p
2	$c^o/c^o \times c^{sr}/c^{sr}$	c^{sr}/c^o	3 $c^{sr}/-$: 1 c^o/c^o
3	$c^o/c^o \times c^p/c^p$	c^o/cc^p	3 $c^o/-$: 1 c^p/c^p

Cross 4 presents a new situation. The color of the F_1 differs from that of either parent, suggesting that two separate genes are involved. An alternative explanation is either codominance or incomplete dominance. If either codominance or incomplete dominance is involved, then the F_2 will appear in a 1:2:1 ratio. If two genes are involved, then a 9:3:3:1 ratio, or some variant of it, will be observed. Because the wild-type phenotype appears in the F_1 and F_2, it appears that complementation is occurring. This requires two genes. The progeny actually are in a 9:4:3 ratio. This means that two genes are involved and that there is epistasis. Furthermore, for three phenotypes to be present in the F_2, the two F_1 parents must have been heterozygous.

Let a stand for the scarlet gene and A for its colorless allele, and assume that there is a dominant allele, C, that blocks the expression of the alleles that we have been studying to this point.

Cross 4: P c^o/c^o ; $A/A \times C/C$; a/a

  F_1 C/c^o ; A/a

  F_2 9 $C/-$; $A/-$ yellow
       3 $C/-$; a/a scarlet
       3 c^o/c^o ; $A/-$ orange
       1 c^o/c^o ; a/a orange (epistasis, with c^o blocking the expression of a/a)

31. a.   P A/A (agouti) $\times a/a$ (nonagouti)
    gametes A and a

    F_1 A/a (agouti)
    gametes A and a

	F$_2$	1 *A/A* (agouti) : 2 *A/a* (agouti) : 1 *a/a* (nonagouti)

b. P   *B/B* (wild type) × *b/b* (cinnamon)
       gametes   *B* and *b*

       F$_1$ *B/b* (wild type)
       gametes   *B* and *b*

       F$_2$ 1 *B/B* (wild type) : 2 *B/b* (wild type) : 1 *b/b* (cinnamon)

c. P   *A/A* ; *b/b* (cinnamon or brown agouti) × *a/a* ; *B/B* (black nonagouti)
       gametes   *A* ; *b* and *a* ; *B*

       F$_1$ *A/a* ; *B/b* (wild type, or black agouti)

d. 9 *A/–* ; *B/–*   black agouti
       3 *a/a* ; *B/–*   black nonagouti
       3 *A/–*; *b/b*   cinnamon
       1 *a/a* ; *b/b*   chocolate

e. P   *A/A* ; *b/b* (cinnamon) × *a/a* ; *B/B* (black nonagouti)
       gametes   *A* ; *b* and *a* ; *B*

       F$_1$ *A/a* ; *B/b* (wild type)
       gametes   *A* ; *B*, *A* ; *b*, *a* ; *B*, and *a* ; *b*

       F$_2$ 9   *A/–* ; *B/–*   wild type
                 1 *A/A* ; *B/B*
                 2 *A/a* ; *B/B*
                 2 *A/A* ; *B/b*
                 4 *A/a* ; *B/b*

            3   *a/a* ; *B/–*   black nonagouti
                 1 *a/a* ; *B/B*
                 2 *a/a* ; *B/b*

            3   *A/–* ; *b/b*   cinnamon
                 1 *A/A* ; *b/b*
                 2 *A/a* ; *b/b*

            1   *a/a* ; *b/b*   chocolate

f. P   *A/a* ; *B/b* × *A/A* ; *b/b*   *A/a* ; *B/b* × *a/a* ; *B/B*
            (wild type) (cinnamon)   (wild type) (black nonagouti)

       F$_1$ 1/4 *A/A* ; *B/b* wild type   1/4 *A/a* ; *B/B* wild type
            1/4 *A/a* ; *B/b* wild type   1/4 *A/a* ; *B/b* wild type

$\frac{1}{4}$ A/A ; b/b cinnamon $\frac{1}{4}$ a/a ; B/B black nonagouti

$\frac{1}{4}$ A/a ; b/b cinnamon $\frac{1}{4}$ a/a ; B/b black nonagouti

g.   P A/a ; $B/b \times a/a$; b/b

         (wild type) (chocolate)

    F_1 $\frac{1}{4}$ A/a ; B/b wild type

         $\frac{1}{4}$ A/a ; b/b cinnamon

         $\frac{1}{4}$ a/a ; B/b black nonagouti

         $\frac{1}{4}$ a/a ; b/b chocolate

h.   To be albino, the mice must be c/c, but the genotype with regard to the A and B genes can be determined only by looking at the F_2 progeny.

Cross 1:   P c/c ; $?/?$; $?/? \times C/C$; A/A ; B/B

           F_1 C/c ; $A/-$; $B/-$

           F_2 87 wild type $C/-$; $A/-$; $B/-$

                      32 cinnamon $C/-$; $A/-$; b/b

                      39 albino c/c ; $?/?$; $?/?$

For cinnamon to appear in the F_2, the F_1 parents must be B/b. Because the wild type is B/B, the albino parent must have been b/b. Now the F_1 parent can be written C/c ; $A/-$; B/b. With such a cross, one-fourth of the progeny would be expected to be albino (c/c), which is what is observed. Three-fourths of the remaining progeny would be black, either agouti or nonagouti, and one-fourth would be either cinnamon, if agouti, or chocolate, if nonagouti. Because chocolate is not observed, the F_1 parent must not carry the allele for nonagouti. Therefore, the F_1 parent is A/A and the original albino must have been c/c ; A/A ; b/b.

Cross 2:   P c/c ; $?/?$; $?/? \times C/C$; A/A ; B/B

           F_1 C/c ; $A/-$; $B/-$

           F_2 62 wild type $C/-$; $A/-$; $B/-$

                      18 albino c/c ; $?/?$; $?/?$

This is a 3:1 ratio, indicating that only one gene is heterozygous in the F_1. That gene must be C/c. Therefore, the albino parent must be c/c ; A/A ; B/B.

Cross 3:   P c/c ; $?/?$; $?/? \times C/C$; A/A ; B/B

F$_1$ C/c ; $A/-$; $B/-$

F$_2$ 96 wild type $C/-$; $A/-$; $B/-$
          30 black $C/-$; a/a ; $B/-$
          41 albino c/c ; $?/?$; $?/?$

For a black nonagouti phenotype to appear in the F$_2$, the F$_1$ must have been heterozygous for the A gene. Therefore, its genotype can be written C/c ; A/a ; $B/-$ and the albino parent must be c/c ; a/a ; $?/?$. Among the colored F$_2$ a 3:1 ratio is observed, indicating that only one of the two genes is heterozygous in the F$_1$. Therefore, the F$_1$ must be C/c ; A/a ; B/B and the albino parent must be c/c ; a/a ; B/B.

Cross 4:   P c/c ; $?/?$; $?/?$ × C/C ; A/A ; B/B

F$_1$ C/c ; $A/-$; $B/-$

F$_2$ 287 wild type $C/-$; $A/-$; $B/-$
          86 black $C/-$; a/a ; $B/-$
          92 cinnamon $C/-$; $A/-$; b/b
          29 chocolate $C/-$; a/a ; b/b
          164 albino c/c ; $?/?$; $?/?$

To get chocolate F$_2$ progeny the F$_1$ parent must be heterozygous for all genes and the albino parent must be c/c ; a/a ; b/b.

32.   To solve this problem, first restate the information.

$A/-$ yellow $A/-$; $R/-$ gray
$R/-$ black a/a ; r/r white

The cross is gray × yellow, or $A/-$; $R/-$ × $A/-$; r/r. The F$_1$ progeny are

3/8 yellow   1/8 black
3/8 gray   1/8 white

For white progeny, both parents must carry an r and an a allele. Now the cross can be rewritten as: A/a ; R/r × A/a ; r/r

33.   a.   The stated cross is

P   single-combed (r/r ; p/p) × walnut-combed (R/R ; P/P)

F$_1$ R/r ; P/p walnut

F$_2$ 9 $R/-$; $P/-$ walnut

$$3 \ r/r \ ; P/- \quad \text{pea}$$
$$3 \ R/- \ ; p/p \quad \text{rose}$$
$$1 \ r/r \ ; p/p \quad \text{single}$$

b. The stated cross is

P   Walnut-combed × rose-combed

and the F_1 progeny are

Phenotypes		Possible genotypes
$3/8$	rose	$R/- \ ; p/p$
$3/8$	walnut	$R/- \ ; P/-$
$1/8$	pea	$r/r \ ; P/-$
$1/8$	single	$r/r \ ; p/p$

The 3 $R/-$: 1 r/r ratio indicates that the parents were heterozygous for the R gene. The 1 $P/-$: 1 p/p ratio indicates a testcross for this gene. Therefore, the parents were $R/r \ ; P/p$ and $R/r \ ; p/p$.

c. The stated cross is

P   walnut-combed × rose-combed

F_1 walnut $(R/- \ ; P/p)$

To get this result, one of the parents must be homozygous R, but both need not be, and the walnut parent must be homozygous P/P.

d. The following genotypes produce the walnut phenotype

$R/R \ ; P/P$, $R/r \ ; P/P$, $R/R \ ; P/p$, $R/r \ ; P/p$

34. Notice that the F_1 shows a difference in phenotype correlated with sex. At least one of the two genes is X-linked. The F_2 ratio suggests independent assortment between the two genes. Because purple is present in the F_1, the parental white-eyed male must have at least one P allele. The presence of white eyes in the F_2 suggests that the F_1 was heterozygous for pigment production, which means that the male also must carry the a allele. A start on the parental genotypes can now be made

P $A/A \ ; P/P \times a/- \ ; P/-$, where – could be either a Y chromosome or a
                                                        second allele.

The question now is, which gene is X-linked? If the A gene is X-linked, the cross is

P $A/A ; p/p \times a/Y ; P/P$

F$_1$ $A/a ; P/p \times A/Y ; P/p$

All F$_2$ females will inherit the A allele from their father. Under this circumstance, no white-eyed females would be observed. Therefore, the A gene cannot be X-linked. The cross is

P $A/A ; p/p \times a/a ; P/Y$

F$_1$ $A/a ; P/p$ purple-eyed females
         $A/a ; p/Y$ red-eyed males

F$_2$ Females   Males
         $^3/_8 A/- ; P/p$ purple $^3/_8 A/- ; P/Y$ purple
         $^3/_8 A/- ; p/p$ red $^3/_8 A/- ; p/Y$ red
         $^1/_8 a/a ; P/p$ white $^1/_8 a/a ; P/Y$ white
         $^1/_8 a/a ; p/p$ white $^1/_8 a/a ; p/Y$ white

35. The results indicate that two genes are involved (modified 9:3:3:1 ratio), with white blocking the expression of color by the other gene. The ratio of white : color is 3:1, indicating that the F$_1$ is heterozygous (W/w). Among colored dogs, the ratio is 3 black : 1 brown, indicating that black is dominant to brown and the F$_1$ is heterozygous (B/b). The original brown dog is $w/w ; b/b$ and the original white dog is $W/W ; B/B$. The F$_1$ progeny are $W/w ; B/b$ and the F$_2$ progeny are

        9 $W/- ; B/-$ white
        3 $w/w ; B/-$ black
        3 $W/- ; b/b$ white
        1 $w/w ; b/b$ brown

36. a. The cross is

        P $td ; su$ (wild type) $\times td^+ ; su^+$ (wild type)

        F$_1$ 1 $td ; su$ wild type
               1 $td ; su^+$ requires tryptophan
               1 $td^+ ; su^+$ wild type
               1 $td^+ ; su$ wild type

     b. 1 tryptophan–dependent : 3 tryptophan–independent

37. P $A/A ; B/B ; C/C ; D/D ; S/S \times a/a ; b/b ; c/c ; d/d ; s/s$

     F$_1$ $A/a ; B/b ; C/c ; D/d ; S/s$

F_2 A/a ; B/b ; C/c ; D/d ; S/s × A/a ; B/b ; C/c ; D/d ; S/s

$(^3/_4\ A/-)(^3/_4\ B/-)(^3/_4\ C/-)(^3/_4\ D/-)(^3/_4\ S/-) = {}^{243}/_{1024}$ agouti

$(^3/_4\ A/-)(^3/_4\ B/-)(^3/_4\ C/-)(^3/_4\ D/-)(^1/_4\ s/s) = {}^{81}/_{1024}$ spotted agouti

$(^3/_4\ A/-)(^3/_4\ B/-)(^3/_4\ C/-)(^1/_4\ d/d)(^3/_4\ S/-) = {}^{81}/_{1024}$ dilute agouti

$(^3/_4\ A/-)(^3/_4\ B/-)(^3/_4\ C/-)(^1/_4\ d/d)(^1/_4\ s/s) = {}^{27}/_{1024}$ dilute spotted agouti

$(^3/_4\ A/-)(^3/_4\ B/-)(^1/_4\ c/c)(^3/_4\ D/-)(^3/_4\ S/-) = {}^{81}/_{1024}$ albino

$(^3/_4\ A/-)(^3/_4\ B/-)(^1/_4\ c/c)(^3/_4\ D/-)(^1/_4\ s/s) = {}^{27}/_{1024}$ albino

$(^3/_4\ A/-)(^3/_4\ B/-)(^1/_4\ c/c)(^1/_4\ d/d)(^3/_4\ S/-) = {}^{27}/_{1024}$ albino

$(^3/_4\ A/-)(^3/_4\ B/-)(^1/_4\ c/c)(^1/_4\ d/d)(^1/_4\ s/s) = {}^9/_{1024}$ albino

$(^3/_4\ A/-)(^1/_4\ b/b)(^3/_4\ C/-)(^3/_4\ D/-)(^3/_4\ S/-) = {}^{81}/_{1024}$ agouti brown

$(^3/_4\ A/-)(^1/_4\ b/b)(^3/_4\ C/-)(^3/_4\ D/-)(^1/_4\ s/s) = {}^{27}/_{1024}$ agouti brown spotted

$(^3/_4\ A/-)(^1/_4\ b/b)(^3/_4\ C/-)(^1/_4\ d/d)(^3/_4\ S/-) = {}^{27}/_{1024}$ dilute agouti brown

$(^3/_4\ A/-)(^1/_4\ b/b)(^3/_4\ C/-)(^1/_4\ d/d)(^1/_4\ s/s) = {}^9/_{1024}$ dilute spotted agouti brown

$(^3/_4\ A/-)(^1/_4\ b/b)(^1/_4\ c/c)(^3/_4\ D/-)(^3/_4\ S/-) = {}^{27}/_{1024}$ albino

$(^3/_4\ A/-)(^1/_4\ b/b)(^1/_4\ c/c)(^3/_4\ D/-)(^1/_4\ s/s) = {}^9/_{1024}$ albino

$(^3/_4\ A/-)(^1/_4\ b/b)(^1/_4\ c/c)(^1/_4\ d/d)(^3/_4\ S/-) = {}^9/_{1024}$ albino

$(^3/_4\ A/-)(^1/_4\ b/b)(^1/_4\ c/c)(^1/_4\ d/d)(^1/_4\ s/s) = {}^3/_{1024}$ albino

$(^1/_4\ a/a)(^3/_4\ B/-)(^3/_4\ C/-)(^3/_4\ D/-)(^3/_4\ S/-) = {}^{81}/_{1024}$ black

$(^1/_4\ a/a)(^3/_4\ B/-)(^3/_4\ C/-)(^3/_4\ D/-)(^1/_4\ s/s) = {}^{27}/_{1024}$ black spotted

$(^1/_4\ a/a)(^3/_4\ B/-)(^3/_4\ C/-)(^1/_4\ d/d)(^3/_4\ S/-) = {}^{27}/_{1024}$ dilute black

$(^1/_4\ a/a)(^3/_4\ B/-)(^3/_4\ C/-)(^1/_4\ d/d)(^1/_4\ s/s) = {}^9/_{1024}$ dilute spotted black

$(^1/_4\ a/a)(^3/_4\ B/-)(^1/_4\ c/c)(^3/_4\ D/-)(^3/_4\ S/-) = {}^{27}/_{1024}$ albino

$(^1/_4\ a/a)(^3/_4\ B/-)(^1/_4\ c/c)(^3/_4\ D/-)(^1/_4\ s/s) = {}^9/_{1024}$ albino

$(^1/_4\ a/a)(^3/_4\ B/-)(^1/_4\ c/c)(^1/_4\ d/d)(^3/_4\ S/-) = {}^9/_{1024}$ albino

$(^1/_4\ a/a)(^3/_4\ B/-)(^1/_4\ c/c)(^1/_4\ d/d)(^1/_4\ s/s) = {}^3/_{1024}$ albino

$(^1/_4\ a/a)(^1/_4\ b/b)(^3/_4\ C/-)(^3/_4\ D/-)(^3/_4\ S/-) = {}^{27}/_{1024}$ brown

$(^1/_4\ a/a)(^1/_4\ b/b)(^3/_4\ C/-)(^3/_4\ D/-)(^1/_4\ s/s) = {}^9/_{1024}$ brown spotted

$(^1/_4\ a/a)(^1/_4\ b/b)(^3/_4\ C/-)(^1/_4\ d/d)(^3/_4\ S/-) = {}^9/_{1024}$ dilute brown

$(^1/_4\ a/a)(^1/_4\ b/b)(^3/_4\ C/-)(^1/_4\ d/d)(^1/_4\ s/s) = {}^3/_{1024}$ dilute spotted brown

$(^1/_4\ a/a)(^1/_4\ b/b)(^1/_4\ c/c)(^3/_4\ D/-)(^3/_4\ S/-) = {}^9/_{1024}$ albino

$(^1/_4\ a/a)(^1/_4\ b/b)(^1/_4\ c/c)(^3/_4\ D/-)(^1/_4\ s/s) = {}^3/_{1024}$ albino

$(^1/_4\ a/a)(^1/_4\ b/b)(^1/_4\ c/c)(^1/_4\ d/d)(^3/_4\ S/-) = {}^3/_{1024}$ albino

$(^1/_4\ a/a)(^1/_4\ b/b)(^1/_4\ c/c)(^1/_4\ d/d)(^1/_4\ s/s) = {}^1/_{1024}$ albino

38.   **a. and b.** The two starting lines are i/i ; D/D ; M/M ; W/W and I/I ; d/d ; m/m ; w/w, and you are seeking i/i ; d/d ; m/m ; w/w. There are many ways to proceed, one of which follows below.

I. i/i ; D/D ; M/M ; W/W × I/I ; d/d ; m/m ; w/w

II. I/i ; D/d ; M/m ; W/w × I/i ; D/d ; M/m ; W/w

III. Select i/i ; d/d ; m/m ; w/w, which has a probability of $(1/4)^4 = 1/256$.

c. and d. In the first cross, all progeny chickens will be of the desired genotype to proceed to the second cross. Therefore, the only problems are to be sure that progeny of both sexes are obtained for the second cross, which is relatively easy, and that enough females are obtained to make the time required for the desired genotype to appear feasible. Because chickens lay one to two eggs a day, the more females who are egg–laying, the faster the desired genotype will be obtained. In addition, it will be necessary to obtain a male and a female of the desired genotype in order to establish a pure breeding line.

Assume that each female will lay two eggs a day, that money is no problem, and that excess males cause no problems. By hatching 200 eggs from the first cross, approximately 100 females will be available for the second cross. These 100 females will produce 200 eggs each day. Thus, in one week a total of 1,400 eggs will be produced. Of these 1,400 eggs, there will be approximately 700 of each sex. At a probability of $1/256$, the desired genotype should be achieved 2.7 times for each sex within that first week.

39. Pedigrees like this are quite common. They indicate lack of penetrance due to epistasis or environmental effects. Individual A must have the dominant autosomal gene, even though she does not express the trait, as both her children are affected.

40. In cross 1, the following can be written immediately

P $M/-$; $D/-$; w/w (dark purple) × m/m ; $?/?$; $?/?$ (white with yellowish spots)

F$_1$ $1/2$ $M/-$; $D/-$; w/w dark purple
       $1/2$ $M/-$; d/d ; w/w light purple

All progeny are colored, indicating that no W allele is present in the parents. Because the progeny are in a 1:1 ratio, only one of the genes in the parents is heterozygous. Also, the light purple progeny, d/d, indicates which gene that is. Therefore, the genotypes must be

P M/M ; D/d ; w/w × m/m ; d/d ; w/w

F$_1$ $1/2$ M/m ; D/d ; w/w
           $1/2$ M/m ; d/d ; w/w

In cross 2, the following can be written immediately

P   *m/m* ; *?/?* ; *?/?* (white with yellowish spots) × *M/–* ; *d/d* ; *w/w* (light purple)

F$_1$ $^1/_2$ *M/–* ; *?/?* ; *W/–*   white with purple spots
          $^1/_4$ *M/–* ; *D/–* ; *w/w*   dark purple
          $^1/_4$ *M/–* ; *d/d* ; *w/w*   light purple

For light and dark purple to appear in a 1:1 ratio among the colored plants, one of the parents must be heterozygous *D/d*. The ratio of white to colored is 1:1, a testcross, so one of the parents is heterozygous *W/w*. All plants are purple, indicating that one parent is homozygous *M/M*. Therefore, the genotypes are

P   *m/m* ; *D/d* ; *W/w* × *M/M* ; *d/d* ; *w/w*

F$_1$ $^1/_2$ *M/m* ; *–/d* ; *W/w*
          $^1/_4$ *M/m* ; *D/d* ; *w/w*
          $^1/_4$ *M/m* ; *d/d* ; *w/w*

41. a. Note that the first two crosses are reciprocal and that the male offspring differ in phenotype between the two crosses. This indicates that the gene is on the X chromosome.

Also note that the F$_1$ females in the first two crosses are sickle. This indicates that sickle is dominant to round. The third cross also indicates that oval is dominant to sickle. Therefore, this is a multiple allelic series with oval > sickle > round.

Let W^o = oval, W^s = sickle, and W^r = round. The three crosses are

Cross 1: W^s/W^s × W^r/Y → W^s/W^r and W^s/Y

Cross 2: W^r/W^r × W^s/Y → W^s/W^r and W^r/Y

Cross 3: W^s/W^s × W^o/Y → W^o/W^s and W^s/Y

b. W^o/W^s × W^r/Y

          $^1/_4$ W^o/W^r female oval
          $^1/_4$ W^s/W^r female sickle
          $^1/_4$ W^o/Y   male oval
          $^1/_4$ W^s/Y   male sickle

42. a. First note that there is a phenotypic difference between males and females, indicating **X** linkage. This means that the male progeny express both alleles of the female parent. A beginning statement of the genotypes could be as follows, where H indicates the gene and the numbers 1, 2, and 3 indicate the variants.

P $H^1/H^2 \times H^3/Y$

F_1 $\frac{1}{4}\, H^1/H^3$ one-banded female
$\frac{1}{4}\, H^3/H^2$ three-banded female
$\frac{1}{4}\, H^1/Y$ one-banded male
$\frac{1}{4}\, H^2/Y$ two-banded male

Because both female F_1 progeny obtain allele H^3 from their father, yet only one has a three-banded pattern, there is obviously a dominance relationship among the alleles. The mother indicates that H^1 is dominant to H^2. The female progeny must be H^1/H^3 and H^3/H^2, and they have a one-banded and three-banded pattern, respectively. H^3 must be dominant to H^2 because the daughter with that combination of alleles has to be the three-banded daughter. In other words, there is a multiple–allelic series with $H^1 > H^3 > H^2$.

b. The cross is

P $H^3/H^2 \times H^1/Y$

F_1 $\frac{1}{4}\, H^1/H^3$ one-banded female
$\frac{1}{4}\, H^1/H^2$ one-banded female
$\frac{1}{4}\, H^2/Y$ two-banded male
$\frac{1}{4}\, H^3/Y$ three-banded male

43. a. The first two crosses indicate that wild-type is dominant to both platinum and aleutian. The third cross indicates that two genes are involved rather than one gene with multiple alleles, because, a 9:3:3:1 ratio is observed.

Let platinum be a, aleutian be b, and wild type be $A/- ; B/-$.

Cross 1:	P	$A/A ; B/B \times a/a ; B/B$	wild-type × platinum
	F_1	$A/a ; B/B$	all wild-type
	F_2	$3\, A/- ; B/B : 1\, a/a ; B/B$	3 wild-type : 1 platinum
Cross 2:	P	$A/A ; B/B \times A/A ; b/b$	wild-type × aleutian
	F_1	$A/A ; B/b$	all wild-type
	F_2	$3\, A/A ; B/- : 1\, A/A ; b/b$	3 wild-type : 1 aleutian

Cross 3:   P a/a ; B/B × A/A ; b/b platinum × aleutian
            F_1 A/a ; B/b all wild-type
            F_2 9 $A/-$; $B/-$ wild-type
                     3 $A/-$; b/b aleutian
                     3 a/a ; $B/-$ platinum
                     1 a/a ; b/b sapphire

b. sapphire × platinum   sapphire × aleutian
P a/a ; b/b × a/a ; B/B a/a ; b/b × A/A ; b/b
F_1 a/a ; B/b platinum A/a ; b/b aleutian
F_2 3 a/a ; $B/-$ platinum   3 $A/-$; b/b aleutian
       1 a/a ; b/b sapphire   1 a/a ; b/b sapphire

44.   a.   The genotypes are

P B/B ; i/i × b/b ; I/I

F_1 B/b ; I/i hairless

F_2 9 $B/-$; $I/-$ hairless
         3 $B/-$; i/i straight
         3 b/b ; $I/-$ hairless
         1 b/b ; i/i bent

b.   In order to solve this problem, first write as much as you can of the progeny genotypes.

4 hairless $-/-$; $I/-$
3 straight $B/-$; i/i
1 bent b/b ; i/i

Each parent must have a b allele and an i allele. The partial genotypes of the parents are

$-/b$; $-/i$ × $-/b$; $-/i$

At least one parent carries the B allele, and at least one parent carries the I allele. Assume for a moment that the first parent carries both. The partial genotypes become

B/b ; I/i × $-/b$; $-/i$

Note that $1/2$ the progeny are hairless. This must come from a I/i × i/i cross. Of those progeny with hair, the ratio is 3:1, which must come from a B/b × B/b cross. The final genotypes are therefore

$$B/b \; ; I/i \; \times \; B/b \; ; i/i$$

45. a. The first question is whether there are two genes or one gene with three alleles. Note that a black × eyeless cross produces black and brown progeny in one instance but black and eyeless progeny in the second instance. Further note that a black × black cross produces brown, a brown × eyeless cross produces brown, and a brown × black cross produces eyeless. The results are confusing enough that the best way to proceed is by trial and error.

There are two possibilities: one gene or two genes.

One gene. Assume one gene for a moment. Let the gene be E and assume the following designations

E^1 = black
E^2 = brown
E^3 = eyeless

If you next assume, based on the various crosses, that black > brown > eyeless, genotypes in the pedigree become

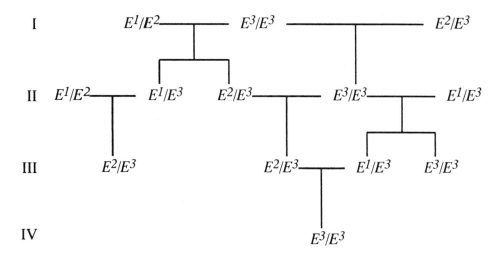

b. The genotype of individual II-3 is E^2/E^3.

Two genes. If two genes are assumed, then questions arise regarding whether both are autosomal or if one might be X-linked. Eye color appears to be autosomal. The presence or absence of eyes appears could be X-linked. Let

B = black X^E = normal eyes
b = brown X^e = eyeless

The pedigree then is

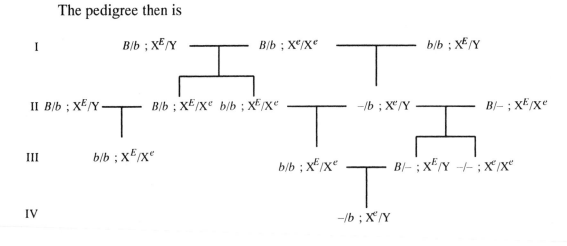

I B/b ; X^E/Y —————— B/b ; X^e/X^e —————— b/b ; X^E/Y

II B/b ; X^E/Y —— B/b ; X^E/X^e b/b ; X^E/X^e —————— $-/b$; X^e/Y —————— $B/-$; X^E/X^e

III b/b ; X^E/X^e b/b ; X^E/X^e —— $B/-$; X^E/Y $-/-$; X^e/X^e

IV $-/b$; X^e/Y

With this interpretation of the pedigree, individual II–3 is b/b ; X^E/X^e.

Without further data, it is impossible to choose scientifically between the two possible explanations.

Questions 46-48 all ask you to compare expected results to actual data to determine if a certain hypothesis is correct. In Chapter 2, you learned how to use a specific analytical test (χ^2) to determine if actual observations deviate from expectations purely on the basis of chance. This would be a good time to review that material to confirm your answers.

46. For the cross $B/b \times B/b$, you expect $3/4$ $B/-$ and $1/4$ b/b or for 400 progeny, 300 and 100 respectively. This should be compared to the actual data of 280 and 120. Alternatively, if the B/B genotype was actually lethal, then you would expect $2/3$ B/b and $1/3$ b/b or for 400 progeny, 267 and 133 respectively. This compares more favorably to the actual data. (The χ^2 value for the 3:1 hypothesis is 5.33 (reject). χ^2 value for the 2:1 hypothesis is 2.253 (accept).)

47. There are a total of 159 progeny that should be distributed in 9:3:3:1 ratio if the two genes are assorting independently. You can see that

Observed	Expected
88 $P/-$; $Q/-$	90
32 $P/-$; q/q	30
25 p/p ; $Q/-$	30
14 p/p ; q/q	10

(For the 9:3:3:1 hypothesis, the χ^2 value is 2.61, accept.)

48. For 160 progeny, there should be 90:70, 130:30, or 120:40 of phenotype 1 : phenotype 2 progeny for the various expected ratios. Only the 9:7 ratio seems feasible and that would indicate that two genes are assorting independently where

$A/- ; B/-$ phenotype 1
$A/- ; b/b$ phenotype 2
$a/a ; B/-$ phenotype 2
$a/a ; b/b$ phenotype 2

(χ^2 values are 2.53, accept, for 90:70 ratio; > 30, reject, for 130 to 30 ratio; and >10, reject, for 120 to 40 ratio.)

49. a. Intercrossing mutant strains that all share a common recessive phenotype is the basis of the complementation test. This test is designed to identify the number of different genes that can mutate to a particular phenotype. If the progeny of a given cross still express the mutant phenotype, the mutations fail to complement and are considered alleles of the same gene; if the progeny are wild type, the mutations complement and the two strains carry mutant alleles of separate genes.

b. There are 3 genes represented in these crosses. All crosses except 2 × 3 (or 3 × 2) complement and indicate that the strains are mutant for separate genes. Strains 2 and 3 fail to complement and are mutant for the same gene.

c. Let A and a represent alleles of gene 1; B and b represent alleles of gene 4; and c^2, c^3, and C represent alleles of gene 3.

Line 1: $a/a \cdot B/B \cdot C/C$
Line 2: $A/A \cdot B/B \cdot c^2/c^2$
Line 3: $A/A \cdot B/B \cdot c^3/c^3$
Line 4: $A/A \cdot b/b \cdot C/C$

	Cross	Genotype	Phenotype
F_1s	1 × 2	$A/a \cdot B/B \cdot C/c^2$	wild type
	1 × 3	$A/a \cdot B/B \cdot C/c^3$	wild type
	1 × 4	$A/a \cdot B/b \cdot C/C$	wild type
	2 × 3	$A/A \cdot B/B \cdot c^2/c^3$	mutant
	2 × 4	$A/A \cdot B/b \cdot C/c^2$	wild type
	3 × 4	$A/A \cdot B/b \cdot C/c^3$	wild type

d. With the exception that strain 2 and 3 fail to complement and therefore have mutations in the same gene, this test does not give evidence of linkage. To test linkage, the F_1s should be crossed to tester strains (homozygous recessive strains) and segregation of the mutant phenotype followed. If the genes are unlinked, for example $A/a ; B/b \times a/a ; b/b$, then 25 percent of the progeny will be wild type ($A/a ; B/b$) and 75 percent will be mutant (25 percent $A/a ; b/b$, 25 percent $a/a ; B/b$, and 25 percent $a/a ; b/b$). If the genes are linked ($a\ B/a\ B \times A\ b/A\ b$) then only one half of the recombinants (i.e., less than 25 percent of the total progeny) will be wild type (*A B/a b*).

e. No. All it tells you is that among these strains, there are three genes represented. If genetic dissection of leg coordination was desired, large screens for the mutant phenotype would be executed with the attempt to "saturate" the genome with mutations in all genes involved in the process.

50. a. Complementation refers to gene products within a cell, which is not what is happening here. Most likely, what is known as cross-feeding is occurring, whereby a product made by one strain diffuses to another strain and allows growth of the second strain. This is equivalent to supplementing the medium. Because cross-feeding seems to be taking place, the suggestion is that the strains are blocked at different points in the metabolic pathway.

b. For cross-feeding to occur, the growing strain must have a block that occurs earlier in the metabolic pathway than does the block in the strain from which it is obtaining the product for growth.

c. The *trpE* strain grows when cross-fed by either *trpD* or *trpB*, but the converse is not true (placing *trpE* earlier in the pathway than either *trpD* or *trpB*), and *trpD* grows when cross-fed by *trpB* (placing *trpD* prior to *trpB*). This suggests that the metabolic pathway is

$$trpE \rightarrow trpD \rightarrow trpB$$

d. Without some tryptophan, no growth at all would occur, and the cells would not have lived long enough to produce a product that could diffuse.

CHALLENGING PROBLEMS

51. Note that the F$_2$ are in an approximate 9:6:1 ratio. This suggests a dihybrid cross in which *A/–* ; *b/b* has the same appearance as *a/a* ; *B/–*. Let the disc phenotype be the result of *A/–* ; *B/–* and the long phenotype be the result of *a/a* ; *b/b*. The crosses are

P *A/A* ; *B/B* (disc) × *a/a* ; *b/b* (long)

F$_1$ *A/a* ; *B/b* (disc)

F$_2$ 9 *A/–* ; *B/–* (disc)
3 *a/a* ; *B/–* (sphere)
3 *A/–* ; *b/b* (sphere)
1 *a/a* ; *b/b* (long)

52. **a.** The best explanation is that Marfan's syndrome is inherited as a dominant autosomal trait, because roughly half of the children of all affected individuals also express the trait. If it were recessive, then all individuals marrying affected spouses would have to be heterozygous for an allele that, when homozygous, causes Marfan's.

b. The pedigree shows both pleiotropy (multiple affected traits) and variable expressivity (variable degree of expressed phenotype). Penetrance is the percentage of individuals with a specific genotype who express the associated phenotype. There is no evidence of decreased penetrance in this pedigree.

c. Pleiotropy indicates that the gene product is required in a number of different tissues, organs, or processes. When the gene is mutant, all tissues needing the gene product will be affected. Variable expressivity of a phenotype for a given genotype indicates modification by one or more other genes, random noise, or environmental effects.

53.

Cross	Results	Conclusion
$A/-$; $C/-$; $R/-$ × a/a ; c/c ; R/R	50% colored	Colored or white will depend on the A and C genes. Because half the seeds are colored, one of the two genes is heterozygous.
$A/-$; $C/-$; $R/-$ × a/a ; C/C ; r/r	25% colored	Color depends on A and R in this cross. If only one gene were heterozygous, 50% would be colored. Therefore, both A and R are heterozygous. The colored plant is A/a ; C/C; R/r.
$A/-$; $C/-$; $R/-$ × A/A ; c/c ; r/r	50% colored	This supports the above conclusion.

54. **a.** The A/A ; C/C ; R/R ; pr/pr parent produces pigment that is not converted to purple. The phenotype is red. The a/a ; c/c ; r/r ; Pr/Pr does not produce pigment. The phenotype is yellow.

b. The F_1 will be A/a ; C/c ; R/r ; Pr/pr, which will produce pigment. The pigment will be converted to purple.

c. The difficult way to determine the phenotypic ratios is to do a branch diagram, yielding the following results.

$^{81}/_{256}$ $A/-$; $C/-$; $R/-$; $Pr/-$	purple		$^{9}/_{256}$ a/a ; $C/-$; $R/-$; $Pr/-$	yellow
$^{27}/_{256}$ $A/-$; $C/-$; $R/-$; pr/pr	red		$^{9}/_{256}$ a/a ; $C/-$; $R/-$; pr/pr	yellow
$^{27}/_{256}$ $A/-$; $C/-$; r/r ; $Pr/-$	yellow		$^{9}/_{256}$ a/a ; $C/-$; r/r ; $Pr/-$	yellow
$^{9}/_{256}$ $A/-$; $C/-$; r/r ; pr/pr	yellow		$^{3}/_{256}$ a/a ; $C/-$; r/r ; pr/pr	yellow
$^{27}/_{256}$ $A/-$; c/c ; $R/-$; $Pr/-$	yellow		$^{27}/_{256}$ a/a ; c/c ; $R/-$; $Pr/-$	yellow
$^{9}/_{256}$ $A/-$; c/c ; $R/-$; pr/pr	yellow		$^{3}/_{256}$ a/a ; c/c ; $R/-$; pr/pr	yellow
$^{9}/_{256}$ $A/-$; c/c ; r/r ; $Pr/-$	yellow		$^{3}/_{256}$ a/a ; c/c ; r/r ; $Pr/-$	yellow
$^{3}/_{256}$ $A/-$; c/c ; r/r ; pr/pr	yellow		$^{1}/_{256}$ a/a ; c/c ; r/r ; pr/pr	yellow

The final phenotypic ratio is 81 purple : 27 red : 148 yellow.

The easier method of determining phenotypic ratios is to recognize that four genes are involved in a heterozygous × heterozygous cross. Purple requires a dominant allele for each gene. The probability of all dominant alleles is $(^3/_4)^4 = {}^{81}/_{256}$. Red results from all dominant alleles except for the *Pr* gene. The probability of that outcome is $(^3/_4)^3(^1/_4) = {}^{27}/_{256}$. The remainder of the outcomes will produce no pigment, resulting in yellow. That probability is $1 - {}^{81}/_{256} - {}^{27}/_{256} = {}^{148}/_{256}$.

d.   The cross is A/a ; C/c ; R/r ; Pr/pr × a/a ; c/c ; r/r ; pr/pr. Again, either a branch diagram or the easier method can be used. The final probabilities are

$$\text{purple} = (^1/_2)^4 = {}^1/_{16}$$
$$\text{red} = (^1/_2)^3(^1/_2) = {}^1/_{16}$$
$$\text{yellow} = 1 - {}^1/_{16} - {}^1/_{16} = {}^7/_8$$

55. **a.**   This type of gene interaction is called *epistasis*. The phenotype of e/e is epistatic to the phenotypes of $B/-$ or b/b.

b.   The progeny of generation I have all possible phenotypes. Progeny II-3 is beige (e/e) so both parents must be heterozygous E/e. Progeny II-4 is brown (b/b) so both parents must also be heterozygous B/b. Progeny III-3 and III-5 are brown so II-2 and II-5 must be B/b. Progeny III-2 and III-7 are beige (e/e) so all their parents must be E/e.

The following are the inferred genotypes

I   1 ($B/b\ E/e$)   2 ($B/b\ E/e$)
II   1 ($b/b\ E/e$)   2 ($B/b\ E/e$)   3 ($-/-\ e/e$)   4 ($b/b\ E/-$)   5 ($B/b\ E/e$)
   6 ($b/b\ E/e$)
III   1 ($B/b\ E/-$)   2 ($-/b\ e/e$)   3 ($b/b,\ E/-$)   4 ($B/b\ E/-$)   5 ($b/b\ E/-$)
   6 ($B/b\ E/-$)   7 ($-/b\ e/e$)

56. a. This is a dihybrid cross resulting in a 13 white : 3 red ratio of progeny in the F_2. This ratio of white to red indicates that the double recessive is not the red phenotype. Instead, the general formula for color is represented by a/a ; $B/–$.

Let line 1 be A/A ; B/B and line 2 be a/a ; b/b. The F_1 is A/a ; B/b. Assume that A blocks color in line 1 and b/b blocks color in line 2. The F_1 will be white because of the presence of A. The F_2 would be

9 $A/–$; $B/–$	white
3 $A/–$; b/b	white
3 a/a ; $B/–$	red
1 a/a ; b/b	white

b. Cross 1: A/A ; $B/B \times A/a$; $B/b \rightarrow$ all $A/–$; $B/–$ white

Cross 2: a/a ; $b/b \times A/a$; B/b

$^1/_4$ A/a ; B/b	white
$^1/_4$ A/a ; b/b	white
$^1/_4$ a/a ; b/b	white
$^1/_4$ a/a ; B/b	red

57. a. Note that blue is always present, indicating E/E (blue) in both parents. Because of the ratios that are observed, neither C nor D is varying. In this case, the gene pairs that are involved are A/a and B/b. The parents are A/A ; $b/b \times a/a$; B/B or A/A ; $B/B \times a/a$; b/b.

The F_1 are A/a ; B/b and the F_2 are

9	$A/–$; $B/–$	blue + red, or purple
3	$A/–$; b/b	blue + yellow, or green
3	a/a ; $B/–$	blue + white$_2$, or blue
1	a/a ; b/b	blue + white$_2$, or blue

b. Blue is not always present, indicating E/e in the F_1. Because green never appears, the F_1 must be B/B ; C/C ; D/D. The parents are A/A ; $e/e \times a/a$; E/E or A/A ; $E/E \times a/a$; e/e.

The F_1 are A/a ; E/e, and the F_2 are

9	$A/–$; $E/–$	red + blue, or purple
3	$A/–$; e/e	red + white$_1$, or red
3	a/a ; $E/–$	white$_2$ + blue, or blue
1	a/a ; e/e	white$_2$ + white$_1$, or white

c. Blue is always present, indicating that the F_1 is E/E. No green appears, indicating that the F_1 is also B/B. The two genes involved are A and D. The parents are $A/A \; ; d/d \; \times \; a/a \; ; D/D$ or $A/A \; ; D/D \; \times \; a/a \; ; d/d$.

The F_1 are $A/a \; ; D/d$ and the F_2 are

9	$A/- \; ; D/-$	blue + red + white$_4$, or purple
3	$A/- \; ; d/d$	blue + red, or purple
3	$a/a \; ; D/-$	blue + white$_2$ + white$_4$, or blue
1	$a/a \; ; d/d$	white$_2$ + blue + red, or purple

d. The presence of yellow indicates $b/b \; ; e/e$ in the F_2. Therefore, the parents are $B/B \; ; e/e \; \times \; b/b \; ; E/E$ or $B/B \; ; E/E \; \times \; b/b \; ; e/e$.

The F_1 are $B/b \; ; E/e$ and the F_2 are

9	$B/- \; ; E/-$	red + blue, or purple
3	$B/- \; ; e/e$	red + white$_1$, or red
3	$b/b \; ; E/-$	yellow + blue, or green
1	$b/b \; ; e/e$	yellow + white$_1$, or yellow

e. Mutations in D suppress mutations in A.

f. Recessive alleles of A will be epistatic to mutations in B.

58. a. Note that cross 1 suggests that one gene is involved and that single is dominant to double. Cross 2 supports this conclusion. Now, note that in crosses 3 and 4, a 1:1 ratio is seen in the progeny, suggesting that superdouble is an allele of both single and double. Superdouble must be heterozygous, however, and it must be dominant to both single and double. Because the heterozygous superdouble yields both single and double when crossed with the appropriate plant, it cannot be heterozygous for the single allele. Therefore, it must be heterozygous for the double allele. A multiple allelic series has been detected: superdouble > single > double.

For now, assume that only one gene is involved and attempt to rationalize the crosses with the assumptions made above.

Cross	Parents	Progeny	Conclusion
1	$A^S/A^S \times A^D/A^D$	A^S/A^D	A^S is dominant to A^D
2	$A^S/A^D \times A^S A^D$	$3 \; A^S/- : 1 \; A^D/A^D$	supports above conclusion

3    $A^D/A^D \times A^{Sd}/A^D$ $1\ A^{Sd}/A^D : 1\ A^D/A^D$ A^{Sd} is dominant to A^D

4    $A^S/A^S \times A^{Sd}/A^D$ $1\ A^{Sd}/A^S : 1\ A^S/A^D$ A^{Sd} is dominant to A^S

5    $A^D/A^D \times A^{Sd}/A^S$ $1\ A^{Sd}/A^D : 1\ A^D/A^S$ supports conclusion of heterozygous superdouble

6    $A^D/A^D \times A^S/A^D$ $1\ A^D/A^D : 1\ A^D/A^S$ supports conclusion of heterozygous superdouble

b. While this explanation does rationalize all the crosses, it does not take into account either the female sterility or the origin of the superdouble plant from a double-flowered variety.

A number of genetic mechanisms could be proposed to explain the origin of superdouble from the double-flowered variety. Most of the mechanisms will be discussed in later chapters and so will not be mentioned here. However, it can be safely assumed at this point that, whatever the mechanism, it was aberrant enough to block the proper formation of the complex structure of the female flower. Because of female sterility, no homozygote for superdouble can be observed.

59. a. All the crosses suggest two independently assorting genes. However, that does not mean that there are a total of only two genes governing eye color. In fact, there are four genes controlling eye color that are being studied here. Let

a/a = defect in the yellow line 1
b/b = defect in the yellow line 2
d/d = defect in the brown line
e/e = defect in the orange line

The genotypes of each line are as follows

yellow 1: $a/a\ ; B/B\ ; D/D\ ; E/E$
yellow 2: $A/A\ ; b/b\ ; D/D\ ; E/E$
brown: $A/A\ ; B/B\ ; d/d\ ; E/E$
orange: $A/A\ ; B/B\ ; D/D\ ; e/e$

b. P $a/a\ ; B/B\ ; D/D\ ; E/E \times A/A\ ; b/b\ ; D/D\ ; E/E$ yellow 1 × yellow 2

F_1 $A/a\ ; B/b\ ; D/D\ ; E/E$ red

F_2 $9\ A/-\ ; B/-\ ; D/D\ ; E/E$ red
$3\ a/a\ ; B/-\ ; D/D\ ; E/E$ yellow
$3\ A/-\ ; b/b\ ; D/D\ ; E/E$ yellow

1 a/a ; b/b ; D/D ; E/E yellow

P a/a ; B/B ; D/D ; E/E × A/A ; B/B ; d/d ; E/E yellow 1 × brown

F_1 A/a ; B/B ; D/d ; E/E red

F_2 9 $A/-$; B/B ; $D/-$; E/E red
       3 a/a ; B/B ; $D/-$; E/E yellow
       3 $A/-$; B/B ; d/d ; E/E brown
       1 a/a ; B/B ; d/d ; E/E yellow

P a/a ; B/B ; D/D ; E/E × A/A ; B/B ; D/D ; e/e yellow 1 × orange

F_1 A/a ; B/B ; D/D ; E/e red

F_2 9 $A/-$; B/B ; D/D ; $E/-$ red
       3 a/a ; B/B ; D/D ; $E/-$ yellow
       3 $A/-$; B/B ; D/D ; e/e orange
       1 a/a ; B/B ; D/D ; e/e orange

P A/A ; b/b ; D/D ; E/E × A/A ; B/B ; d/d ; E/E yellow 2 × brown

F_1 A/A ; B/b ; D/d ; E/E red

F_2 9 A/A ; $B/-$; $D/-$; E/E red
       3 A/A ; b/b ; $D/-$; E/E yellow
       3 A/A ; $B/-$; d/d ; E/E brown
       1 A/A ; b/b ; d/d ; E/E yellow

P A/A ; b/b ; D/D ; E/E × A/A ; B/B ; D/D ; e/e yellow 2 × orange

F_1 A/A ; B/b ; D/D ; E/e red

F_2 9 A/A ; $B/-$; D/D ; $E/-$ red
       3 A/A ; b/b ; D/D ; $E/-$ yellow
       3 A/A ; $B/-$; D/D ; e/e orange
       1 A/A ; b/b ; D/D ; e/e yellow

P A/A ; B/B ; d/d ; E/E × A/A ; B/B ; D/D ; e/e brown × orange

F_1 A/A ; B/B ; D/d ; E/e red

F_2 9 A/A ; B/B ; $D/-$; $E/-$ red
       3 A/A ; B/B ; d/d ; $E/-$ brown
       3 A/A ; B/B ; $D/-$; e/e orange
       1 A/A ; B/B ; d/d ; e/e orange

c. When constructing a biochemical pathway, remember that the earliest gene that is defective in a pathway will determine the phenotype of a doubly defective genotype. Look at the following doubly recessive homozygotes from the crosses. Notice that the double defect d/d ; e/e has the same phenotype as the defect a/a ; e/e. This suggests that the E gene functions earlier than do the A and D genes. Using this logic, the following table can be constructed

Genotype	Phenotype	Conclusion
a/a ; B/B ; D/D ; e/e	orange	E functions before A
A/A ; B/B ; d/d ; e/e	orange	E functions before D
a/a ; b/b ; D/D ; E/E	yellow	B functions before A
A/A ; b/b ; D/D ; e/e	yellow	B functions before E
a/a ; B/B ; d/d ; E/E	yellow	A functions before D
A/A ; b/b ; d/d ; E/E	yellow	B functions before D

The genes function in the following sequence: B, E, A, D. The metabolic path is

$$\text{yellow 2} \quad \underset{B}{\rightarrow} \quad \text{orange} \quad \underset{E}{\rightarrow} \quad \text{yellow} \quad \underset{A}{\rightarrow} \quad \text{brown} \quad \underset{D}{\rightarrow} \quad \text{red}$$

60. a. A trihybrid cross would give a 63:1 ratio. Therefore, there are three R loci segregating in this cross.

b.

P R_1/R_1 ; R_2/R_2 ; R_3/R_3 × r_1/r_1 ; r_2/r_2 ; r_3/r_3

F₁ R_1/r_1 ; R_2/r_2 ; R_3/r_3

F₂

27	$R_1/-$; $R_2/-$; $R_3/-$	red
9	$R_1/-$; $R_2/-$; r_3/r_3	red
9	$R_1/-$; r_2/r_2 ; $R_3/-$	red
9	r_1/r_1 ; $R_2/-$; $R_3/-$	red
3	$R_1/-$; r_2/r_2 ; r_3/r_3	red
3	r_1/r_1 ; $R_2/-$; r_3/r_3	red
3	r_1/r_1 ; r_2/r_2 ; $R_3/-$	red
1	r_1/r_1 ; r_2/r_2 ; r_3/r_3	white

c. (1) In order to obtain a 1:1 ratio, only one of the genes can be heterozygous. A representative cross would be R_1/r_1 ; r_2/r_2 ; r_3/r_3 × r_1/r_1 ; r_2/r_2 ; r_3/r_3.

(2) In order to obtain a 3 red : 1 white ratio, two alleles must be segregating and they cannot be within the same gene. A representative cross would be $R_1/r_1 ; R_2/r_2 ; r_3/r_3 \times r_1/r_1 ; r_2/r_2 ; r_3/r_3$.

(3) In order to obtain a 7 red : 1 white ratio, three alleles must be segregating, and they cannot be within the same gene. The cross would be $R_1/r_1 ; R_2/r_2 ; R_3/r_3 \times r_1/r_1 ; r_2/r_2 ; r_3/r_3$.

    d.   The formula is $1 - (1/4)^n$, where n = the number of loci that are segregating in the representative crosses above.

61. **a.**   The trait is recessive (parents without the trait have children with the trait) and autosomal (daughters can inherit the trait from unaffected fathers). Looking to generation III, there is also evidence that there are two different genes that, when defective, result in deaf-mutism.

        Assuming that one gene has alleles A and a, and the other has B and b. The following genotypes can be inferred:

I-1 and I-2	$A/a ; B/B$	I-3 and I-4	$A/A ; B/b$
II-(1, 3, 4, 5, 6)	$A/- ; B/B$	II- (9, 10, 12, 13, 14, 15)	$A/A ; B/-$
II-2 and II-7	$a/a ; B/B$	I-8 and II-11	$A/A ; b/b$

    b.   Generation III shows complementation. All are $A/a ; B/b$.

62. **a.**   The first impression from the pedigree is that the gene causing blue sclera and brittle bones is pleiotropic with variable expressivity. If two genes were involved, it would be highly unlikely that all people with brittle bones also had blue sclera.

    b.   Sons and daughters inherit from affected fathers so the allele appears to be autosomal.

    c.   The trait appears to be inherited as a dominant but with incomplete penetrance. For the trait to be recessive, many of the non-related individuals marrying into the pedigree would have to be heterozygous (e.g., I-1, I-3, II-8, II-11). Individuals II-4, II-14, III-2, and III-14 have descendants with the disorder although they do not themselves express the disorder. Therefore, $4/20$ people that can be inferred to carry the gene, do not express the trait. That is 80 percent penetrance. (Penetrance could be significantly less than that since many possible carriers have no shown progeny.) The pedigree also exhibits variable expressivity. Of the 16 individuals who have blue sclera, 10 do not have brittle bones. Usually, expressivity is put in terms of none, variable, and highly variable, rather than expressed as percentages.

63. **a. and b.** Assuming that both the *Brown* and the *Van Scoy* lines were homozygous, the parental cross suggests that nonhygienic behavior is dominant to hygienic. A consideration of the specific behavior of the F_1 × *Brown* progeny, however, suggests that the behavior is separable into two processes; one involves uncapping and one involves removal of dead pupae. If this is true, uncapping and removal of dead pupae, two behaviors that normally go together in the *Brown* line, have been separated in the F_2 progeny, suggesting that they are controlled by different and unlinked genes. Those bees that lack uncapping behavior are still able to express removal of dead pupae if environmental conditions are such that they do not need to uncap a compartment first. Also, the uncapping behavior is epistatic to removal of dead pupae.

Let U = no uncapping, u = uncapping, R = no removal, r = removal

P $\quad u/u\,;r/r \times U/U\,;R/R$
$\qquad$ (*Brown*) (*Van Scoy*)

$\qquad U/u\,;R/r \times u/u\,;r/r$ (F_1 cross *Brown*)

Progeny $\quad 1/4\;U/u\,;R/r \quad$ nonhygienic
$\qquad\qquad 1/4\;U/u\,;r/r \quad$ nonhygienic (no uncapping but removal if uncapped)
$\qquad\qquad 1/4\;u/u\,;R/r \quad$ uncapping but no removal
$\qquad\qquad 1/4\;u/u\,;r/r \quad$ hygienic

64. **a. and b.** Cross 1 indicates that orange is dominant to yellow. Crosses 2–4 indicate that red is dominant to orange, yellow, and white. Crosses 5–7 indicate that there are two genes involved in the production of color. Cross 5 indicates that yellow and white are different genes. Cross 6 indicates that orange and white are different genes.

In other words, epistasis is involved and the homozygous recessive white genotype seems to block production of color by a second gene.

Begin by explaining the crosses in the simplest manner possible until such time as it becomes necessary to add complexity. Therefore, assume that orange and yellow are alleles of gene *A* with orange dominant.

Cross 1: $\quad$ P $\qquad A/A \times a/a$
$\qquad\qquad$ F_1 $\qquad A/a$
$\qquad\qquad$ F_2 $\qquad 3\;A/- : 1\;a/a$

Immediately there is trouble in trying to write the genotypes for cross 2, unless red is a third allele of the same gene. Assume the following dominance relationships: red > orange > yellow. Let the alleles be designated as follows

red	A^R
orange	A^O
yellow	A^Y

Crosses 1-3 now become

P	$A^O/A^O \times A^Y/A^Y$	$A^R/A^R \times A^O/A^O$	$A^R/A^R \times A^Y/A^Y$
F1	A^O/A^Y	A^R/A^O	A^R/A^Y
F2	$3\,A^O/- : 1\,A^Y/A^Y$	$3\,A^R/- : 1\,A^O/A^O$	$3\,A^R/- : 1\,A^Y/A^Y$

Cross 4: To do this cross you must add a second gene. You must also rewrite the above crosses to include the second gene. Let B allow color and b block color expression, producing white. The first three crosses become

P	$A^O/A^O ; B/B \times A^Y/A^Y ; B/B$	$A^R/A^R ; B/B \times A^O/A^O ; B/B$	$A^R/A^R ; B/B \times A^Y/A^Y ; B/B$
F_1	$A^O/A^Y ; B/B$	$A^R/A^O ; B/B$	$A^R/A^Y ; B/B$
F_2	$3\,A^O/- ; B/B : 1\,A^Y/A^Y ; B/B$	$3\,A^R/- ; B/B : 1\,A^O/A^O ; B/B$	$3\,A^R/- : 1\,A^Y/A^Y ; B/B$

The fourth cross is

P	$A^R/A^R ; B/B \times A^R/A^R ; b/b$
F_1	$A^R/A^R ; B/b$
F_2	$3\,A^R/A^R ; B/- : 1\,A^R/A^R ; b/b$

Cross 5: To do this cross, note that there is no orange appearing. Therefore, the two parents must carry the alleles for red and yellow, and the expression of red must be blocked.

P	$A^Y/A^Y ; B/B \times A^R/A^R ; b/b$	
F_1	$A^R/A^Y ; B/b$	
F_2	$9\,A^R/- ; B/-$	red
	$3\,A^R/- ; b/b$	white

$$3\ A^Y/A^Y\ ;\ B/-\qquad\text{yellow}$$
$$1\ A^Y/A^Y\ ;\ b/b\qquad\text{white}$$

Cross 6:   This cross is identical with cross 5 except that orange replaces yellow.

P $\qquad\qquad A^O/A^O\ ;\ B/B\ \times\ A^R/A^R\ ;\ b/b$

F$_1$ $\qquad\qquad A^R/A^O\ ;\ B/b$

F$_2$ $\qquad\qquad$

$9\ A^R/-\ ;\ B/-$	red
$3\ A^R/-\ ;\ b/b$	white
$3\ A^O/A^O\ ;\ B/-$	orange
$1\ A^O/A^O\ ;\ b/b$	white

Cross 7:   In this cross, yellow is suppressed by b/b.

P $\qquad\qquad A^R/A^R\ ;\ B/B\ \times\ A^Y/A^Y\ ;\ b/b$

F$_1$ $\qquad\qquad A^R/A^Y\ ;\ B/b$

F$_2$ $\qquad\qquad$

$9\ A^R/-\ ;\ B/-$	red
$3\ A^R/-\ ;\ b/b$	white
$3\ A^Y/A^Y\ ;\ B/-$	yellow
$1\ A^Y/A^Y\ ;\ b/b$	white

65. The ratios observed in the F$_2$ of these crosses are all variations of 9:3:3:1, indicating that two genes are assorting and epistasis is occurring.

(1)   F$_1$ $A/a\ ;\ B/b\ \times\ A/a\ ;\ B/b$

F$_2$

$9\ A/-\ ;\ B/-$	cream
$3\ A/-\ ;\ b/b$	cream
$3\ a/a\ ;\ B/-$	black
$1\ a/a\ ;\ b/b$	gray

A test cross of the F$_1$ would be $A/a\ ;\ B/b\ \times\ a/a\ ;\ b/b$ and the offspring:

$1/4\ A/a\ ;\ B/b$	cream
$1/4\ a/a\ ;\ B/b$	black
$1/4\ A/a\ ;\ b/b$	cream
$1/4\ a/a\ ;\ b/b$	gray

(2) F_1 A/a ; B/b × A/a ; B/b

         F_2 9 $A/-$; $B/-$ orange
                   3 $A/-$; b/b yellow
                   3 a/a ; $B/-$ yellow
                   1 a/a ; b/b yellow

A test cross of the F_1 would be A/a ; B/b × a/a ; b/b and the offspring:

     $1/4$ A/a ; B/b orange
     $1/4$ a/a ; B/b yellow
     $1/4$ A/a ; b/b yellow
     $1/4$ a/a ; b/b yellow

(3) F_1 A/a ; B/b × A/a ; B/b

         F_2 9 $A/-$; $B/-$ black
                   3 $A/-$; b/b white
                   3 a/a ; $B/-$ black
                   1 a/a ; b/b black

A test cross of the F_1 would be A/a ; B/b × a/a ; b/b and the offspring:

     $1/4$ A/a ; B/b black
     $1/4$ a/a ; B/b black
     $1/4$ A/a ; b/b white
     $1/4$ a/a ; b/b black

(4) F_1 A/a ; B/b × A/a ; B/b

         F_2 9 $A/-$; $B/-$ solid red
                   3 $A/-$; b/b mottled red
                   3 a/a ; $B/-$ small red dots
                   1 a/a ; b/b small red dots

A test cross of the F_1 would be A/a ; B/b × a/a ; b/b and the offspring:

     $1/4$ A/a ; B/b solid red
     $1/4$ a/a ; B/b small red dots
     $1/4$ A/a ; b/b mottled red
     $1/4$ a/a ; b/b small red dots

66. a. Intercrossing mutant strains that all share a common recessive phenotype is the basis of the complementation test. This test is designed to identify the number of different genes that can mutate to a particular phenotype. In this problem, if the progeny of a given cross still express the wiggle phenotype, the mutations fail to complement and are considered alleles of the same gene; if the progeny are wild type, the mutations complement and the two strains carry mutant alleles of separate genes.

b. From the data,

1 and 5	fail to complement	gene A
2, 6, 8, and 10	fail to complement	gene B
3 and 4	fail to complement	gene C
7, 11, and 12	fail to complement	gene D
9	complements all others	gene E

There are five complementation groups (genes) identified by this data.

c. mutant 1: $a^1/a^1 \cdot b^+/b^+ \cdot c^+/c^+ \cdot d^+/d^+ \cdot e^+/e^+$ (although, only the mutant alleles are usually listed)

mutant 2: $a^+/a^+ \cdot b^2/b^2 \cdot c^+/c^+ \cdot d^+/d^+ \cdot e^+/e^+$

mutant 5: $a^5/a^5 \cdot b/^+b^+ \cdot c^+/c^+ \cdot d^+/d^+ \cdot e^+/e^+$

1/5 hybrid: $a^1/a^5 \cdot b^+/b^+ \cdot c^+/c^+ \cdot d^+/d^+ \cdot e^+/e^+$ phenotype: wiggles

Conclusion: 1 and 5 are both mutant for gene A

2/5 hybrid: $a^+/a^5 ,\cdot b^+/b^2 \cdot c^+/c^+ \cdot d^+/d^+ \cdot e^+/e^+$ phenotype: wild type

Conclusion: 2 and 5 are mutant for different genes

67. Assume that the two genes are unlinked. Mossy would be genotypically m^- ; s^+ and spider would be m^+ ; s^-. After mating and sporulation, independent assortment would give three types of tetrads. The doubly mutant spores do not germinate. Because either single mutant is slow-growing, the additive effect of both mutant genes is lethal.

A		B		C	
m^+ ; s^+	wild type	m^+ ; s^+	wild type	m^+ ; s^-	spider
m^+ ; s^+	wild type	m^+ ; s^-	spider	m^+ ; s^-	spider
m^- ; s^-	dead	m^- ; s^+	mossy	m^- ; s^+	mossy
m^- ; s^-	dead	m^- ; s^-	dead	m^- ; s^+	mossy

7

DNA: Structure and Replication

BASIC PROBLEMS

1. The DNA double helix is held together by two types of bonds, covalent and hydrogen. Covalent bonds occur within each linear strand and strongly bond the bases, sugars, and phosphate groups (both within each component and between components). Hydrogen bonds occur between the two strands and involve a base from one strand with a base from the second in complementary pairing. These hydrogen bonds are individually weak but collectively quite strong.

2. Conservative replication is a hypothetical form of DNA synthesis in which the two template strands remain together but dictate the synthesis of two new DNA strands, which then form a second DNA helix. The end point is two double helices, one containing only old DNA and one containing only new DNA. This hypothesis was found to be not correct. Semiconservative replication is a form of DNA synthesis in which the two template strands separate and each dictates the synthesis of a new strand. The end point is two double helices, both containing one new and one old strand of DNA. This hypothesis was found to be correct.

3. A primer is a short segment of RNA that is synthesized by primase using DNA as a template during DNA replication. Once the primer is synthesized, DNA polymerase then adds DNA to the 3′ end of the RNA. Primers are required because the major DNA polymerase involved with DNA replication is unable to initiate DNA synthesis and, rather, requires a 3′ end. (It is the 3′ -OH group that is required to create the next phosphodiester bond.) The RNA is subsequently removed and replaced with DNA so that no gaps exist in the final product.

4. Helicases are enzymes that disrupt the hydrogen bonds that hold the two DNA strands together in a double helix. This breakage is required for both RNA and DNA synthesis. Topoisomerases are enzymes that create and relax supercoiling in the DNA double helix. The supercoiling itself is a result of the twisting of the DNA helix that occurs when the two strands separate.

5. Because the DNA polymerase is capable of adding new nucleotides only at the 3′ end of a DNA strand, and because the two strands are antiparallel, at least two molecules of DNA polymerase must be involved in the replication of any specific region of DNA. When a region becomes single-stranded, the two strands have an opposite orientation. Imagine a single-stranded region that runs from right to left. The 5′ end is at the right, with the 3′ end pointing to the left; synthesis can initiate and continue uninterrupted toward the right end of this strand. Remember: new nucleotides are added in a 5′→ 3′ direction, so the template must be copied from its 3′ end. The other strand has a 5′ end at the left with the 3′ end pointing right. Thus, the two strands are oriented in opposite directions (antiparallel), and synthesis (which is 5′→ 3′) must proceed in opposite directions. For the leading strand (say, the top strand) replication is to the right, following the replication fork. It is continuous and may be thought of as moving "downstream." Replication on the bottom strand cannot move in the direction of the fork (to the right) because, for this strand, that would mean adding nucleotides to its 5′ end. Therefore, this strand must replicate discontinuously: as the fork creates a new single-stranded stretch of DNA, this is replicated *to the left* (away from the direction of fork movement). For this lagging strand, the replication fork is always opening new single-stranded DNA for replication *upstream* of the previously replicated stretch, and a new fragment of DNA is replicated back to the previously created fragment. Thus, one (Okazaki) fragment follows the other in the direction of the replication fork, but each fragment is created in the opposite direction.

6. No. The information of DNA is dependent on a faithful copying mechanism. The strict rules of complementarity ensure that replication and transcription are reproducible.

7. Helicases are enzymes that disrupt the hydrogen bonds that hold the two DNA strands together in a double helix. This breakage exposes lengths of single-stranded DNA that will act as the template and are required for DNA replication. Therefore, the absence of helicases would prevent the replication process.

8. Theoretically, DNA could be replicated this way but not with the replisome, which is organized to replicate both stands simultaneously. Further, this would leave one strand of the DNA single-stranded where mutagenic events would be more likely, and it would certainly take longer.

9. The chromosome would become hopelessly fragmented.

10. c. DNA synthesis might take longer.

11. d. It does not have to adapt to each of the body's tissues.

12. b. The RNA would be more likely to contain errors.

13. Part of the replisome is the sliding clamp, which encircles the DNA and keeps pol III attached to the DNA molecule. Thus pol III is transformed into a processive enzyme capable of adding tens of thousands of nucleotides.

14. **d.** Replication would take twice as long.

15. **a.** Prior to the S phase, each chromosome has two telomeres, so in the case of $2n = 14$, there are 14 chromosomes and 28 telomeres.

    b. After S, each chromosome consists of two chromatids, each with two telomeres, for a total of four telomeres per chromosome. So, for 14 chromosomes, there would be $14 \times 4 = 56$ telomeres.

    c. At prophase, the chromosomes still consist of two chromatids each, so there would be $14 \times 4 = 56$ telomeres.

    d. At telophase, there would be 28 telomeres in each of the soon-to-be daughter cells.

16. If the DNA is double stranded, A = T and G = C and A + T + C + G = 100%. If T = 15%, then C = $[100 - 15(2)]/2 = 35\%$.

17. If the DNA is double stranded, G = C = 24% and A = T = 26% .

18. Six. The first replication start would have two replication forks proceeding to completion, and the now replicated origins would each start replication again. Each would have two more replication forks, for a total of six.

19. Only the DNA molecule that used the poly-T strand as a template would be radioactive. The other daughter molecule would not be radioactive, because it would not have required any dATP for its replication.

Because each strand of the second molecule contains T, both daughter molecules would require dATP for replication, so each would be radioactive.

20. Yes. DNA replication is also semi-conservative in diploid eukaryotes.

21. **a.** The bottom strand.

b.

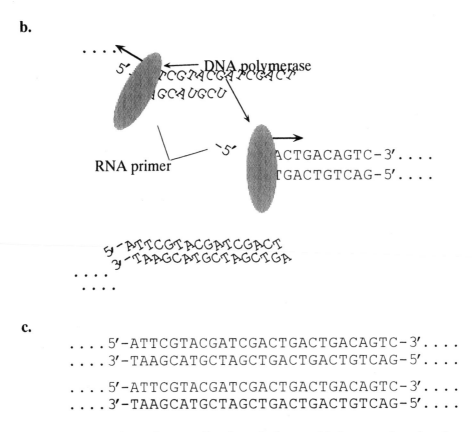

c.

```
.... 5′-ATTCGTACGATCGACTGACTGACAGTC-3′....
.... 3′-TAAGCATGCTAGCTGACTGACTGTCAG-5′....

.... 5′-ATTCGTACGATCGACTGACTGACAGTC-3′....
.... 3′-TAAGCATGCTAGCTGACTGACTGTCAG-5′....
```

d. Yes, but the other replication fork would be moving in the opposite direction and the top strand, as drawn, would now be the leading strand and the bottom strand would now be the lagging strand.

22. The bottom strand will serve as the template for the Okazaki fragment so its sequence will be:

5′....CCTTAAGACTAACTACTTACTGGGATC....3′

23. a.

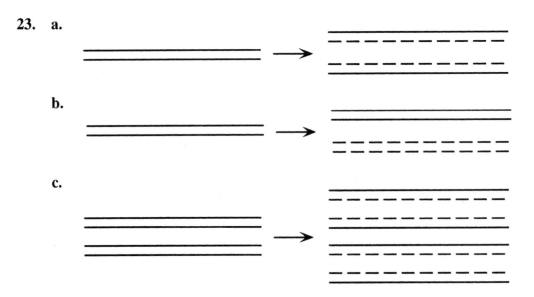

b.

c.

d.

e. Model (b) is ruled out by the experiment. The results are compatible with semiconservative replication, but the exact structure could not be predicted. The other models would all give one band of intermediate density.

24. Sample A must be live S cells because it kills mice when injected and S cells are recovered from the mice.

    Sample C must be live R cells because it has no effect on mice when injected but live R cells are recovered.

    Sample B must be DNA from S cells as it transforms sample C when co-injected.

25. If all samples remain the same, only the line will change when samples B + C are co-injected. Because B contains DNA, it will not transform the R cells, so injected mice will have no response and live R cells will be recovered.

CHALLENGING PROBLEMS

26. Without functional telomerase, the telomeres would shorten at each replication cycle, leading to eventual loss of essential coding information and death. In fact, there are some current observations that decline or loss of telomerase activity plays a role in the mechanism of aging in humans.

27. **a.** A very plausible model is of a triple helix, which would look like a braid, with each strand interacting by hydrogen bonding to the other two.
    b. Replication would have to be terti-conservative. The three strands would separate, and each strand would dictate the synthesis of the other two strands.
    c. The reductional division would have to result in three daughter cells, and the equational would have to result in two daughter cells, in either order. Thus, meiosis would yield six gametes.

28. Chargaff's rules are that A = T and G = C. Because this is not observed, the most likely interpretation is that the DNA is single-stranded. The phage would first have to synthesize a complementary strand before it could begin to make multiple copies of itself.

8

RNA: Transcription and Processing

BASIC PROBLEMS

1. Because RNA can hybridize to both strands, the RNA must be transcribed from both strands. This does not mean, however, that both strands are used as a template *within each gene*. The expectation is that only one strand is used within a gene but that different genes are transcribed in different directions along the DNA. The most direct test would be to purify a specific RNA coding for a specific protein and then hybridize it to the λ genome. Only one strand should hybridize to the purified RNA.

2. In prokaryotes, translation is beginning at the 5′ end while the 3′ end is still being transcribed. In eukaryotes, processing (capping, splicing) is occurring at the 5′ end while the 3′ end is still being transcribed.

3. There are many examples of proteins that act on nucleic acids, but some mentioned in this chapter are RNA polymerase, GTFs (general transcription factors), σ (sigma factor), rho, TBP (TATA binding protein), and snRNPs (a combination of proteins and snRNAs).

4. Sigma factor, as part of the RNA polyermase holoenzyme, recognizes and binds to the −35 and −10 regions of bacterial promoters. It positions the holoenzyme to correctly initiate transcription at the start site. In eukaryotes, TBP (TATA binding protein) and other GTFs (general transcription factors) have an analagous function.

5. The CTD (carboxy tail domain) of the ß subunit of RNA polymerase II contains binding sites for enzymes and other proteins that are required for RNA processing and is located near the site where nascent RNA emerges. If mutations in this subunit prevent the correct binding and/or localization of the proteins necessary for capping, then this modification will not occur even though all the required enzymes are normal.

6. For a given gene, only one strand of DNA is transcribed. This strand (called the template) will be complementary to the RNA and also to the other strand (called the nontemplate or coding strand). Consequently, the nucleotide sequence of the

RNA must be the same as that of the nontemplate strand of the DNA (except that the Ts are instead Us). Ultimately, it is the nucleotide sequence that gives the RNA its function. Transcription of both strands would give two complementary RNAs that would code for completely different polypeptides. Also, double-stranded RNA initiates cellular processes that lead to its degradation.

7.   **a., b., c., and d.**

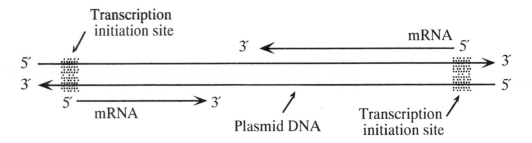

8.   Yes. Both replication and transcription is performed by large, multi-subunit molecular machines (the replisome and RNA polymerase II, respectively) and both require helicase activity at the fork of the bubble. However, transcription proceeds in only one direction and only one DNA strand is copied.

9.   **a.**   False. Sigma factor is required in prokaryotes, not eukaryotes.

     b.   True. Processing begins at the 5′ end, while the 3′ end is still being synthesized.

     c.   False. Processing occurs in the nucleus and only mature RNA is transported out to the cytoplasm.

     d.   False. Hairpin loops or rho factor (in conjunction with the *rut* site) is used to terminate transcription in prokaryotes. In eukaryotes the conserved sequences AAUAAA or AUUAAA, near the 3′ end of the transcript is recognized by an enzyme that cuts off the end of the RNA approximately 20 bases downstream.

     e.   True. Multiple RNA polymerases may transcribe the same template simultaneously.

10. **a.**   The original sequence represents the −35 and −10 consensus sequences (with the correct number of intervening spaces) of a bacterial promoter. Sigma factor, as part of the RNA polymerase holoenzyme, recognizes and binds to these sequences.

     b.   The mutated (transposed) sequences would not be a binding site for sigma factor. The two regions are not in the correct orientation to each other and therefore would not be recognized as a promoter.

11. **a.** The promoters of eukaryotes and prokaryotes do not have the same conserved sequences. In yeast, the promoter would have the required TATA box located about –30 whereas bacteria would have conserved sequences at –35 and –10 that would interact with sigma factor as part of the RNA polymerase holoenzyme.

b. There are two possible reasons that the mRNA is longer than expected. First, many eukaryotic genes contain introns, and bacteria would not have the splicing machinery necessary for their removal. Second, termination of transcription is not the same in bacteria and yeast; the sequences necessary for correct terminaton in *E. coli* would not be expected in the yeast gene.

12.

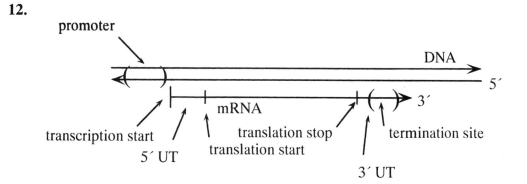

13.

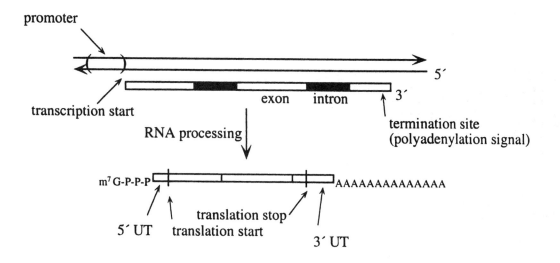

14. **a.** Yes. The exons encode the protein, so null mutations would be expected to map within exons.

b. Possibly. There are sequences near the boundaries of and within introns that are necessary for correct splicing. If these are altered by mutation, correct splicing will be disrupted. Although transcribed, it is likely that translation will not occur.

    c.   Yes. If the promoter is deleted or altered such that GTFs cannot bind, transcription will be disrupted.

    d.   Yes. There are sequences near the boundaries of and within introns that are necessary for correct splicing.

15.   Self-splicing introns are capable of excising themselves from a primary transcript without the need of additional enzymes or energy source. They are one of many examples of RNA molecules that are catalytic, and for this property, they are also known as ribozymes. With this additional function, RNA is the only known biological molecule to encode genetic information and catalyze biological reactions. In simplest terms, it is possible that life began with an RNA molecule, or group of molecules, that evolved the ability to self-replicate.

16.   Antibiotics need to selectively target bacterial structures and functions that are essential for life but unique or sufficiently different from the equivalent structure and functions of their animal hosts. Bacterial RNA polymerase fits these criteria as its function is obviously essential, yet its structure is sufficiently different from the several eukaryotic RNA polymerases. These differences make it possible to develop drugs that specifically bind bacterial RNA polymerase but have little or no affinity for eukaryotic RNA polymerases.

CHALLENGING PROBLEMS

17. a.   The data cannot indicate whether one or both strands are used for transcription. You do not know how much of the DNA is transcribed nor which regions of DNA are transcribed. Only when the purine/pyrimidine ratio is not unity can you deduce that only one strand is used as template.

    b.   If the RNA is double-stranded, the percentage of purines (A + G) would equal the percentage of pyrimidines (U + C) and the (A +G)/(U + C) ratio would be 1.0. This is clearly not the case for *E. coli*, which has a ratio of 0.80. The ratio for *B. subtilis* is 1.02. This is consistent with the RNA being double-stranded but does not rule out single-stranded if there are an equal number of purines and pyrimidines in the strand.

18. a.

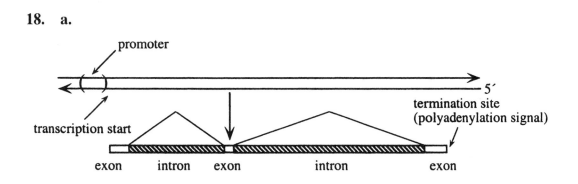

b. Alternative splicing of the primary transcript would result in mRNAs that were only partially identical. In this case, the two transcripts share 224 nucleotides in common. As this is the exact length of the second exon, one possible solution to this problem is that this exon is shared by the two alternatively-spliced mRNAs. The second transcript also contains 2.3 kb of sequence not found in the first. Perhaps what was considered the first intron is actually also part of the second transcript as that would result in the 2524 nucleotides stated as this transcript's length. Of course, other combinations of alternative splicing would also fit the data.

19. Double-stranded RNA, composed of a sense strand and a complementary antisense strand, can be used in *C. elegans* (and likely all organisms) to selectively prevent the synthesis of the encoded gene product (a discovery awarded the 2006 Nobel Prize in Medicine). This process, called gene silencing, blocks the synthesis of the encoded protein from the endogenous gene and is thus equivalent to "knocking out" the gene. To test whether a specific mRNA encodes an essential embryonic protein, eggs or very early embryos should be injected with the double-stranded RNA produced from your mRNA, thus activating the RNAi pathway. The effects of knocking out the specific gene product can then be followed by observing what happens in these, versus control, embryos. If the encoded protein is essential, embryonic development should be perturbed when your gene is silenced.

20. Transgene silencing is a common phenomenon in plants. Silencing may occur at the transcriptional or post-transcriptional level. Since you cannot control where transgenes insert, some may insert into transcriptionally inactive parts of the genome. Post-transcriptional silencing may be the result of activation of the RNAi pathway due to the misexpression of both strands of your transgene. See Figure 8-22 in the companion text as one example of how this can happen. In these cases, the RNAi pathway will be activated and the EPSPS gene product will be silenced.

21. RNAi has the potential to selectively prevent protein production from any targeted gene. Oncogenes are mutant versions of "normal" genes (called proto-oncogenes) and their altered gene products are partly or wholly responsible for causing cancer. In theory, it may be possible to design appropriate siRNA molecules (small interfering RNAs) that specifically silence the mutant oncogene product but do not silence the closely related proto-oncogene product. The latter is necessary to prevent serious side effects as the products of proto-oncogenes are essential for normal cellular function.

9

Proteins and Their Synthesis

Basic Problems

1. | | | | | |
    |---|---|---|---|---|
    | 3´ CGT | ACC | ACT | GCA 5´ | DNA double helix (transcribed strand) |
    | 5´ GCA | TGG | TGA | CGT 3´ | DNA double helix |
    | 5´ GCA | UGG | UGA | CGU 3´ | mRNA transcribed |
    | 3´ CGU | ACC | ACU | GCA 5´ | appropriate tRNA anticodon |

    NH_3 - Ala - Trp - (stop) - COOH   amino acids incorporated

2. **a. and b.**   5´UUG GGA AGC 3´

    c. and d.   Assuming the reading frame starts at the first base:

    NH_3 - Leu - Gly - Ser - COOH

    For the bottom strand, the mRNA is 5´ GCU UCC CAA 3´ and assuming the reading frame starts at the first base, the corresponding amino acid chain is NH_3 - Ala - Ser - Gln - COOH.

3. (5) With an insertion, the reading frame is disrupted. This will result in a drastically altered protein from the insertion to the end of the protein (which may be much shorter or longer than wild type because of the location of stop signals in the altered reading frame).

4. A single nucleotide change should result in three adjacent amino acid changes in a protein. One and two adjacent amino acid changes would be expected to be much rarer than the three changes. This is directly the opposite of what is observed in proteins. Also, given any triplet coding for an amino acid, the next triplet could only be one of four. For example, if the first is GGG, then the next must be GGN (N = any base). This puts severe limits on which amino acids could be adjacent to each other. You could check amino acid sequences of various proteins to show that this is not the case.

5. It suggests very little evolutionary change between *E. coli* and humans with regard to the translational apparatus. The code is universal, the ribosomes are interchangeable, the tRNAs are interchangeable, and the enzymes involved are interchangeable. (Initiation of translation in prokaryotes *in vivo* requires specific base-pairing between the 3′ end of the 16s rRNA and a Shine–Delgano sequence found in the 5′ untranslated region of the mRNA. A Shine–Delgano sequence would not be expected (unless by chance) in a eukaryotic mRNA and therefore initiation of translation might not occur.)

6. There are three codons for isoleucine: 5′ AUU 3′, 5′ AUC 3′, and 5′ AUA 3′. Possible anticodons are 3′ UAA 5′ (complementary), 3′ UAG 5′ (complementary), and 3′ UAI 5′ (wobble). 5′ AAU 3′, although complementary, would also base-pair with 5′ AUG 3′ (methionine) due to wobble and therefore would not be an acceptable alternative.

7. **a.** By studying the genetic code table provided in the textbook, you will discover that there then there are 28 codons that do not specify a particular amino acid with the first two positions (32 if you count Tyr and the stop codons starting with UA).

   b. If you knew the amino acid, you would not know the first two nucleotides in the cases of Arg, Ser, and Leu.

8. The codon for amber is UAG. Listed below are the amino acids that would have been needed to be inserted to continue the wild-type chain and their codons:

glutamine	CAA, CAG*
lysine	AAA, AAG*
glutamic acid	GAA, GAG*
tyrosine	UAU*, UAC*
tryptophan	UGG*
serine	AGU, AGC, UCU, UCC, UCA, UCG*

   In each case, the codon marked by an asterisk would require a single base change to become UAG.

9. **a.** The codons for phenylalanine are UUU and UUC. Only the UUU codon can exist with randomly positioned A and U. Therefore, the chance of UUU is $(1/2)(1/2)(1/2) = 1/8$.

   b. The codons for isoleucine are AUU, AUC, and AUA. AUC cannot exist. The probability of AUU is $(1/2)(1/2)(1/2) = 1/8$, and the probability of AUA is $(1/2)(1/2)(1/2) = 1/8$. The total probability is thus $1/4$.

   c. The codons for leucine are UUA, UUG, CUU, CUC, CUA, and CUG, of which only UUA can exist. It has a probability of $(1/2)(1/2)(1/2) = 1/8$.

d. The codons for tyrosine are UAU and UAC, of which only UAU can exist. It has a probability of $(1/2)(1/2)(1/2) = 1/8$.

10. a. 1 U : 5 C — The probability of a U is $1/6$, and the probability of a C is $5/6$.

Codon	Amino acid	Probability	Sum
UUU	Phe	$(1/6)(1/6)(1/6) = 0.005$	Phe = 0.028
UUC	Phe	$(1/6)(1/6)(5/6) = 0.023$	
CCC	Pro	$(5/6)(5/6)(5/6) = 0.578$	Pro = 0.694
CCU	Pro	$(5/6)(5/6)(1/6) = 0.116$	
UCC	Ser	$(1/6)(5/6)(5/6) = 0.116$	Ser = 0.139
UCU	Ser	$(1/6)(5/6)(1/6) = 0.023$	
CUC	Leu	$(5/6)(1/6)(5/6) = 0.116$	Leu = 0.139
CUU	Leu	$(5/6)(1/6)(1/6) = 0.023$	

1 Phe : 25 Pro : 5 Ser : 5 Leu

b. Using the same method as above, the final answer is 4 stop : 80 Phe : 40 Leu : 24 Ile : 24 Ser : 20 Tyr : 6 Pro : 6 Thr : 5 Asn : 5 His : 1 Lys : 1 Gln.

c. All amino acids are found in the proportions seen in the code table.

11. Quaternary structure is due to the interactions of subunits of a protein. In this example, the enzyme activity being studied may be from a protein consisting of two different subunits. Both subunits are required for activity. The polypeptides of the subunits are encoded by separate and unlinked genes.

12. There are a number of mutational changes that can lead to the absence of enzymatic function in the product of a gene. Some of these changes would result in the complete absence of protein product and therefore also the absence of a detectable band on a Western blot. Mutations such as deletions of the gene, for example, would result in the lack of detectable protein. Other mutations that destroy function (missense, for example) may not alter the production of the protein and would be detected on a Western blot. Still other mutations (nonsense, frameshift) could alter the size of the protein yet would still lead to detectable protein.

13. a. Tryptophan synthetase is a heterotetramer of two copies each of two different polypeptides, each encoded by a separate gene. Mutations that prevent the synthesis of one subunit would lead to the loss of one of the bands on a Western blot. Mutants that still make both subunits (those with exactly the same bands as wild type) might have mutations that prevent the subunits from interacting or disrupt the active site of the enzyme.

b. Because the two subunits are encoded by separate genes, the absence of both bands simultaneously would require two independent and rare mutagenic events.

14. Yes. It was not known at the time what number of bases the "plus" and "minus" mutations actually were. If each mutation was two bases, then a codon would have been six bases. Since the mutations were actually adding or subtracting single bases, the codon is indeed three bases.

15. No. The enzyme may require post-translational modification to be active. Mutations in the enzymes required for these modifications would not map to the isocitrate lyase gene.

16. A nonsense suppressor is a mutation in a tRNA such that its anticodon can base-pair with a stop codon. In this way, a mutant stop codon (nonsense mutation) can be read through and the polypeptide can be fully synthesized. However, the mutant tRNA may be for an amino acid that was not encoded in that position in the original gene. For example, the codon UCG (serine) is instead UAG in the nonsense mutant. The suppressor mutation could be in tRNA for tryptophan such that its anticodon now recognizes UAG instead of UGG. During translation in the double mutant, the machinery puts tryptophan into the location of the mutant stop codon. This allows translation to continue but does place tryptophan into a position that was serine in the wild-type gene. This may create a protein that is not as active and a cell that is "not exactly wild type." Another explanation is that translation of the mutant gene is not as efficient and that premature termination still occurs some of the time. This would lead to less product and again, a state that is "not exactly wild type."

17.

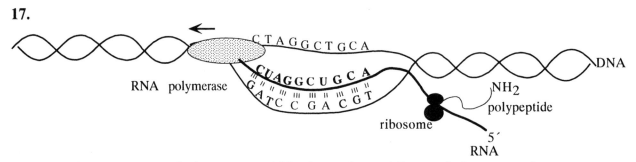

In eukaryotes, transcription occurs within the nucleus while translation occurs in the cytoplasm. Thus the two processes cannot occur together.

18. Assuming that the three mutations of gene P are all nonsense mutations, there are three different possible stop codons that might be the cause (amber, ochre, or opal). A suppressor mutation would be specific to one type of nonsense codon. For example, amber suppressors would suppress amber mutants but not opal or ochre.

19. Initiation of translation in prokaryotes requires specific base-pairing between the 3′ end of the 16s rRNA and a Shine–Delgano sequence found in the 5′ untranslated region of the mRNA. A Shine–Delgano sequence would not be expected (unless by chance) in a eukaryotic mRNA and therefore initiation of translation would not occur.

20. Initiaton of translation in eukaryotes requires initiation factors (eIF4a, b, and G) that associate with the 5′ cap of the mRNA. Because prokaryotic mRNAs are not capped, translation would not initiate.

21. Not likely. Although the steps of translation and the components of ribosomes are similar in both eukaryotes and prokaryotes, the ribosomes are not identical. The sizes of both subunits are larger in eukaryotes and the many specific and intricate interactions that must take place between the small and large subunits would not be possible in a chimeric system.

22. Single amino acid changes can result in changes in protein folding, protein targeting, or post-translational modifications. Any of these changes could give the results indicated.

23. The first indication of rRNAs importance was the discovery of ribozymes. Recently, structural studies have shown that both the decoding center in the 30S subunit and the peptidyl transferase center in the 50S subunit are composed entirely of rRNA and that the important contacts in these centers are all tRNA/rRNA contacts.

24. Antibiotics need to selectively target bacterial structures and functions that are essential for life but unique or sufficiently different from the equivalent structure and functions of their animal hosts. The large bacterial ribosomal subunit fits these criteria as its function is obviously essential yet its structure is sufficiently different from the large eukaryotic ribosomal subunit. While the steps of protein synthesis are similar overall, eukaryotic ribosomes have larger and more numerous components. These differences make it possible to develop drugs that specifically bind bacterial ribosomes but have little or no affinity for eukaryotic ribosomes.

25. Recent studies indicate that most proteins function by interacting with other proteins. (The complete set of such interactions is called the interactome.) Most of these essential protein-protein interactions are regulated by phosphorylation/dephosphorylation modifications. Kinases are the enzymes that catalyze phosphorylations. Therefore, the complexity of the interactome necessary for the complexity of multicellularity requires the very large number of kinase-encoding genes.

26. Bacterial and human cell cultures are both capable of producing the same polypeptide from the same mRNA, but that does not mean that the resulting protein will be active. Many proteins require posttranslational processing to become functional, and the enzymes and control necessary for such processing is not universal. The proteins produced and isolated from a human cell culture system will contain the necessary posttranslational modifications necessary for human protein function.

CHALLENGING PROBLEMS

27. **a. and b.** The goal of this type of problem is to align the two sequences. You are told that there is a single nucleotide addition and single nucleotide deletion, so look for single base differences that effect this alignment. These should be located where the protein sequence changes (i.e., between Lys-Ser and Asn-Ala). Remember also that the genetic code is redundant. (N = any base)

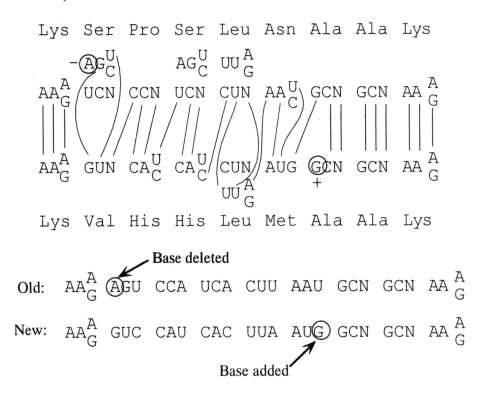

28. Mutant 1: A simple substitution of Arg for Ser exists, suggesting a nucleotide change. Two codons for Arg are AGA and AGG, and one codon for Ser is AGU. The U of the Ser codon could have been replaced by either an A or a G.

Mutant 2: The Trp codon (UGG) changed to a stop codon (UGA or UAG).

Mutant 3: Two frameshift mutations occurred:

5′—GCN CCN (–U)GGA GUG AAA AA(+U or C) UGU(or C) CAU(or C)—3′.

Mutant 4: An inversion occurred after Trp and before Cys. The DNA original sequence (with both strands shown for the area of inversion) was

```
3´—CGN GGN ACC TCA CTT TTT ACA(or G) GTA(or G)—5´
5´—              —AGT GAA AAA—                  —3´
```

Therefore, the complementary RNA sequence was

```
5´—GCN CCN UGG AGU GAA AAA UGU/C CAU/C—3´
```

The DNA inverted sequence became

```
3´—CGN GGN ACC AAA AAG TGA ACA/G GTA/G—5´
            ^         ^
```

Therefore the complementary RNA sequence was

```
5´—GCN CCN UGG UUU UUC ACU UGU/C CAU/C—3´
            ^         ^
```

29. If the anticodon on a tRNA molecule also was altered by mutation to be four bases long, with the fourth base on the 5´ side of the anticodon, it would suppress the insertion. Alterations in the ribosome can also induce frameshifting.

30. f, d, j, e, c, i, b, h, a, g

31. Cells in long-established culture lines usually are not fully diploid. For reasons that are currently unknown, adaptation to culture frequently results in both karyotypic and gene dosage changes. This can result in hemizygosity for some genes, which allows for the expression of previously hidden recessive alleles.

32. **a. and b.** The sequence of double-stranded DNA is as follows:

```
5´—TAC ATG ATC ATT TCA CGG AAT TTC TAG CAT GTA—3´
3´—ATG TAC TAG TAA AGT GCC TTA AAG ATC GTA CAT—5´
```

First look for stop codons. Next look for the initiating codon, AUG (3´—TAC—5´ in DNA). Only the upper strand contains the necessary codons.

```
DNA       3´ TAC GAT CTT TAA GGC ACT 5´
RNA       5´ AUG CUA GAA AUU CCG UGA 3´
protein      Met Leu Glu Ile Pro stop
```

The DNA strand is read from right to left as written in your text and is written above in reverse order from your text.

c. Remember that polarity must be taken into account. The inversion is

```
DNA 5´ TAC ATG CTA GAA ATT CCG TGA AAT GAT CAT GTA 3´
RNA 3´        —GAU CUU UAA GGC ACU UUA CUA GUA—      5´
amino acids     HOOC  7   6   5   4   3   2   1 —NH3
```

d.
```
DNA   3´ATG TAC TAG TAA AGT GCC TTA AAG ATC GTA CAT 5´
mRNA  5´UAC AUG AUC AUU UCA CGG AAU UUC UAG 3´
              1   2   3   4   5   6   7  stop
```

Codon 4 is 5´—UCA—3´, which codes for Ser. Anticodon 4 would be 3´—AGU—5´ (or 3´—AGI—5´ given wobble).

33. **a.** $(GAU)_n$ codes for Asp_n $(GAU)_n$, Met_n $(AUG)_n$, and $stop_n$ $(UGA)_n$. $(GUA)_n$ codes for Val_n $(GUA)_n$, Ser_n $(AGU)_n$, and $stop_n$ $(UAG)_n$. One reading frame in each contains a stop codon.

b. Each of the three reading frames contains a stop codon.

c. The way to approach this problem is to focus initially on one amino acid at a time. For instance, line 4 indicates that the codon for Arg might be AGA or GAG. Line 7 indicates it might be AAG, AGA, or GAA. Therefore, Arg is at least AGA. That also means that Glu is GAG (line 4). Lys and Glu can be AAG or GAA (line 7). Because no other combinations except the ones already mentioned result in either Lys or Glu, no further decision can be made with respect to them. However, taking wobble into consideration, Glu may also be GAA, which leaves Lys as AAG.

Next, focus on lines 1 and 5. Ser and Leu can be UCU and CUC. Ser, Leu, and Phe can be UUC, UCU, and CUU. Phe is not UCU, which is seen in both lines. From line 14, CUU is Leu. Therefore, UUC is Phe, and UCU is Ser.

The footnote states that line 13 and 14 are in the correct order. In line 13, if UCU is Ser (see above), then Ile is AUC, Tyr is UAU, and Leu is CUA.

Continued application of this approach will allow the assignment of an amino acid to each codon.

10

Regulation of Gene Expression in Bacteria and Their Viruses

BASIC PROBLEMS

1. The I gene determines the synthesis of a repressor molecule, which blocks expression of the *lac* operon and which is inactivated by the inducer. The presence of the repressor I^+ will be dominant to the absence of a repressor I^-. I^s mutants are unresponsive to an inducer. For this reason, the gene product cannot be stopped from interacting with the operator and blocking the *lac* operon. Therefore, I^s is dominant to I^+.

2. O^c mutants are changes in the DNA sequence of the operator that impair the binding of the *lac* repressor. Therefore, the *lac* operon associated with the O^c operator cannot be turned off. Because an operator controls only the genes on the same DNA strand, it is cis (on the same strand) and dominant (cannot be turned off).

3. a. You are told that a, b, and c represent *lacI*, *lacO*, and *lacZ*, but you do not know which is which. Both a^- and c^- have constitutive phenotypes (lines 1 and 2) and therefore must represent mutations in either the operator (*lacO*) or the repressor (*lac I*). b^- (line 3) shows no ß-gal activity and by elimination must represent the *lacZ* gene.

    Mutations in the operator will be cis-dominant and will cause constitutive expression of the *lacZ* gene only if it's on the same chromosome. Line 6 has c^- on the same chromosome as b^+ but the phenotype is still inducible (owing to c^+ in trans). Line 7 has a^- on the same chromosome as b^+ and is constitutive even though the other chromosome is a^+. Therefore a is *lacO*, c is *lacI*, and b is *lacZ*.

b. Another way of labeling mutants of the operator is to denote that they lead to a constitutive phenotype; *lacO⁻* (or *a⁻*) can also be written as *lacOᶜ*. There are also mutations of the repressor that fail to bind inducer (allolactose) as opposed to fail to bind DNA. These two classes have quite different phenotypes and are distinguished by *lacIˢ* (fails to bind allolactose and leads to a dominant uninducible phenotype in the presence of a wild-type operator) and *lacI⁻* (fails to bind DNA and is recessive). It is possible that line 3, line 4, and line 7 have *lacIˢ* mutations (because dominance cannot be ascertained in a cell that is also *lacOᶜ*) but the other *c⁻* alleles must be *lacI⁻*.

4.

	ß-Galactosidase		Permease	
Part	No lactose	Lactose	No lactose	Lactose
a	+	+	−	+
b	+	+	−	−
c	−	−	−	−
d	−	−	−	−
e	+	+	+	+
f	+	+	−	−
g	−	+	−	+

a. The *Oᶜ* mutation leads to the constitutive synthesis of ß-galactosidase because it is cis to a *lacZ⁺* gene, but the permease is inducible because the *lacY⁺* gene is cis to a wild-type operator.

b. The *lacP⁻* mutation prevents transcription so only the genes cis to *lacP⁺* will be transcribed. These genes are also cis to *Oᶜ* so the *lacZ⁺* gene is transcribed constitutively.

c. The *lacIˢ* is a trans-dominant mutation and prevents transcription from either operon.

d. Same as part c.

e. There is no functional repressor made (and one operator is mutant as well).

f. Same as part b.

g. Both operators are wild type and the one functional copy of *lacI* will direct the synthesis of enough repressor to control both operons.

5. A gene is turned off or inactivated by the "modulator" (usually called a *repressor*) in negative control, and the repressor must be removed for transcription to occur. A gene is turned on by the "modulator" (usually called an *activator*) in positive control, and the activator must be added or converted to an active form for transcription to occur.

6. The *lacY* gene produces a permease that transports lactose into the cell. A cell containing a *lacY⁻* mutation cannot transport lactose into the cell, so ß-galactosidase will not be induced. (In wild-type cells, even when the repressor is present, there is still a small amount of transcription. This allows a small amount of baseline permease (and β-gal) expression. It is this low level of permease that allows trace amounts of lactose to enter the cell and initiate induction of the *lac* operon.)

7. Both the *lac* operon and phase λ genetic switches can be interpreted as a simple switch between two states. For λ, there is competition between Cro and cI proteins binding at an operator to control the choice between the lysogenic and lytic life cycles. For the *lac* operon, a repressor binds at the operator to prevent transcription of the structural genes in the absence of lactose. Both also are capable of interpreting their environments and directing an "appropriate" response. The *lac* operon responds to the presence or absence of lactose and glucose. For λ, if resources are abundant, or the bacterial cell sustains DNA damage, the lytic cycle prevails, but if resources are not abundant, λ will enter the lysogenic cycle.

8. In the *lac* operon, the promoter is located between the CAP binding site and the repressor binding site (the operator). The *lac* repressor binds at the operator and blocks transcription, and CAP binds at its binding site and activates transcription. Complete induction of the lac structural genes requires the binding of CAP and the absence of binding of repressor. The λ genetic switch consists of an operator (with three binding sites for λ repressor (cI) and Cro) that overlaps two promoters, P_R and P_{RM}. P_R promotes transcription of lytic genes and P_{RM} promotes transcription of the *cI* gene. When repressor is bound, transcription from P_R is blocked, but when Cro is bound, transcription from P_{RM} is blocked.

CHALLENGING PROBLEMS

9. Normally, the repressor searches for the operator by rapidly binding and dissociating from nonoperator sequences. Even for sequences that mimic the true operator, the dissociation time is only a few seconds or less. Therefore, it is easy for the repressor to find new operators as new strands of DNA are synthesized. However, when the affinity of the repressor for DNA and operator is increased, it takes too long for the repressor to dissociate from sequences on the chromosome that mimic the true operator, and as the cell divides and new operators are synthesized, the repressor never quite finds all of them in time, leading to a partial synthesis of ß-galactosidase. This explains why, in the absence of IPTG, there is some elevated ß-galactosidase synthesis. When IPTG

binds to the repressors with increased affinity, it lowers the affinity back to that of the normal repressor (without IPTG bound). Then, the repressor can rapidly dissociate from sequences in the chromosome that mimic the operator and find the true operator. Thus, ß-galactosidase is repressed in the presence of IPTG in strains with repressors that have greatly increased affinity for operator. In summary, because of a kinetic phenomenon, we see a reverse induction curve.

10. If there is an operon governing both genes, then a frameshift mutation could cause the stop codon separating the two genes to be read as a sense codon. Therefore, the second gene product will be incorrect for almost all amino acids. However, there are no known polycistronic messages in eukaryotes. The alternative, and better, explanation is that both enzymatic functions are performed by the same gene product. Here, a frameshift mutation beyond the first function, carbamyl phosphate synthetase, will result in the second half of the protein molecule being nonfunctional.

11. Because very small amounts of the repressor are made, the system as a whole is quite responsive to changes in repressor concentration. In the heterodiploids, repressor heterotetramers may form by association of polypeptides encoded by both I^- and I^+. Only homotetramers of I^+ work properly and their concentration will be reduced by the various heterotetramer combinations of I^- and I^+. This will result in some expression of the *lac* genes in the absence of lactose.

12. The key to this question is to remember that *lacI* mutations will be trans-acting as they produce a protein product, and that *lacP* mutations (like *lacO* mutations) will be cis-acting as they are affecting a binding site for RNA polymerase. For the purposes of this problem, designate the uninducible *lacI* mutations as i^u mutations, and the uninducible *lacP* mutations as p^u mutations. There are quite a number of satisfactory genotypes which can serve to distinguish between the i^u and p^u mutations. Here are a few examples

| | Enzyme activity | |
Genotype	ß-galactosidase	Permease
1 $i^u\ p^+\ o^+\ z^+\ y^+$	absent	absent
2 $i^+\ p^u\ o^+\ z^+\ y^+$	absent	absent
3 $i^u\ p^+\ o^+\ z^+\ y^-\ /\ i^+\ p^+\ o^+\ z^-\ y^+$	absent	absent
4 $i^+\ p^u\ o^+\ z^+\ y^-\ /\ i^+\ p^+\ o^+\ z^-\ y^+$	absent	inducible
5 $i^u\ p^+\ o^c\ z^+\ y^-\ /\ i^+\ p^+\ o^+\ z^-\ y^+$	constitutive	absent
6 $i^+\ p^u\ o^c\ z^+\ y^-\ /\ i^+\ p^+\ o^+\ z^-\ y^+$	absent	inducible

Genotypes 1 and 2 are simply symbolic restatements of the phenotypes of the uninducible mutations. Genotypes 3 and 4 are straightforward tests to distinguish the cis-acting p^u mutations from the trans-acting i^u lesions. The results of genotype 3 reflect the expectation that i^u mutations would be trans-acting and dominant to i^+. This is expected because the i^u-encoded repressor protein molecules would be incapable of being inactivated by binding to inducer; the presence or absence of normal repressor protein is irrelevant. The results of genotype 4, on the other hand, reflect the expectation that p^u mutations would only be cis-acting. Hence, any genes in cis to the p^u allele would be inactive, while any genes in cis to the normal p^+ allele would potentially be transcribed normally (if all other regulatory functions were normal). In a similar fashion, genotypes 5 and 6 distinguish the cis vs. trans action of i^u and p^u mutations. In genotype 5, i^u remains trans-dominant to i^+, but this dominance is overcome by the cis-acting o^c mutation (compare genotypes 3 and 5). In genotype 6, the presence of o^c is irrelevant, as it is in a *lac* operon that contains the p^u mutation preventing RNA polymerase binding (compare genotypes 4 and 6).

13. The S mutation is an alteration in *lacI* such that the repressor protein binds to the operator, regardless of whether inducer is present or not. In other words, it is a mutation that inactivates the allosteric site which binds to inducer, while not affecting the ability of the repressor to bind to the operator site. The dominance of the S mutation is due to the binding of the mutant repressor, even under circumstances when normal repressor does not bind to DNA (that is, in the presence of inducer). The constitutive reverse mutations that map to *lacI* are mutational events which inactivate the ability of this repressor to bind to the operator. The constitutive reverse mutations that map to the operator alter the operator DNA sequence such that it will not permit binding to any repressor molecules (wild-type or mutant repressor).

14. a.

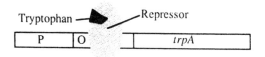

b. i. *trpA* is not synthesized in the presence of tryptophan; it is synthesized in the absence of tryptophan.

   ii. *trpA* is not synthesized in the presence of tryptophan; it is synthesized in the absence of tryptophan. The second operon contains a *trpA⁻* mutation and is unable to make any active tryptophan synthetase enzyme. However, this operon contains a wild-type *trpR* gene, which encodes a functional repressor molecule. Because this gene product is a diffusible molecule (trans-acting), it will act on the other DNA molecule containing the mutant *trpR⁻* gene and bind to the operator on that molecule. Thus, *trpA* gene expression will be repressed when tryptophan binds the repressor, causing the repressor to bind the operator. In the absence of tryptophan, the repressor will not bind to either operator, and transcription will proceed.

iii. *trpA* is synthesized both in the presence and in the absence of tryptophan. Because the second operon contains a mutant *trpA⁻* gene, no functional tryptophan synthetase enzyme will be made from this DNA molecule. The first operon contains a wild-type *trpA* gene and will be responsible for the intracellular supply of tryptophan synthetase and ultimately, tryptophan. However, this operon contains a mutant (*trpO⁻*) operator region that cannot be bound by the repressor molecules. Because the operator region is cis-acting and does not encode a diffusible gene product, the wild-type *trpO* gene on the other DNA molecule cannot substitute for the mutant operator. Therefore, the *trpA* gene will always be transcribed (constitutive) regardless of the levels of tryptophan in the cell.

15.

	Glucose	Lactose	Lactose + glucose
wild-type	0	100	1
lacI⁻	1	100	1
lacIˢ	0	0	0
lacO⁻	1	100	1
crp⁻	0	1	1

lacI⁻ — leads to absence of negative control by repressor binding but still under the positive control of CAP-cAMP binding

lacIˢ — because the repressor does not bind lactose but still binds DNA, transcription will be blocked under all conditions

lacO⁻ — leads to absence of negative control by repressor binding but still under the positive control of CAP-cAMP binding

crp⁻ — leads to absence of positive control by CAP-cAMP binding but still under the negative control of repressor binding

16. Mutations in *cI, cII, cIII* would all effect lysogeny. *cI* encodes the repressor, *cII* encodes an activator of P_{RE}, and *cIII* encodes a protein that protects *cII* from degradation. Mutations in *N* (an antiterminator) would also effect lysogeny as its function is required for transcription of the *cII* and *cIII* genes, but it is also necessary for genes involved in lysis. Mutations in the gene encoding the integrase (*int*) would also effect the ability of a mutant phage to lysogenize.

17. A mutation that prevents Cro binding to O_R would prevent the lytic cycle. *cI* binding would still allow lysogeny. Cro binding is necessary to block the maintenance of *cI* transcription and allow transcription of genes for the lytic cycle.

18. **a.** Sporulation-specific σ factors control many genes. For example, σ^E binds to at least 121 promoters, within 34 operons and 87 individual genes, to regulate more than 250 genes. A loss of this σ factor would prevent the proper transcription of all of these more than 250 genes. The loss of any one of the 121 promoters controlled by this σ factor would only affect the operon or individual gene of that promoter.

b. If you look carefully at the −35 and −10 sequences shown in Figure 10-30b, you will observe that there are differences in these sequences among the genes controlled by the same σ factor. For example, the −35 region of the promoter for the *ybaN* gene is TTATATT, but the same region for the *ydcA* gene is TACTATT, yet both still function correctly. When comparing all promoters regulated by a particular σ factor, a "consensus" sequence can be determined, but there will be variations observed as well. Given this, you would not expect all point mutations in this region to effect gene expression. Some variation is possible.

11

Regulation of Gene Expression in Eukaryotes

BASIC PROBLEMS

1. Activation of gene expression by trans-acting factors occurs in both prokaryotes and eukaryotes. In both cases, the trans-acting factors interact with specific DNA sequences that control expression of cis genes.

   In prokaryotes, proteins bind to specific DNA sequences, which in turn regulate one or more downstream genes.

   In eukaryotes, highly conserved sequences such as CCAAT and various enhancers in conjunction with trans-acting binding proteins increase transcription controlled by the downstream TATA box promoter. Several proteins have been found that bind to the CCAAT sequence, upstream GC boxes, and the TATA sequence in *Drosophila*, yeast, and other organisms. Specifically, the Sp1 protein recognizes the upstream GC boxes of the SV40 promoter and many other genes; GCN4 and GAL4 proteins recognize upstream sequences in yeast; and many hormone receptors bind to specific sites on the DNA (e.g., estrogen complexed to its receptor binding to a sequence upstream of the ovalbumin gene in chicken oviduct cells). Additionally, the structure of some of these trans-acting DNA-binding proteins is quite similar to the structure of binding proteins seen in prokaryotes. Further, protein-protein interactions are important in both prokaryotes and eukaryotes. For the above reasons, eukaryotic regulation is now thought to be very close to the model for regulation of the bacterial *ara* operon.

2. In general, the ground state of a bacterial gene is "on." Thus, transcription initiation is prevented or reduced if the binding of RNA polymerase is blocked. In contrast, the ground state in eukaryotes is "off." Thus, the transcriptional machinery (including RNA polymerase II and associated general transcription factors) cannot bind to the promoter in the absence of other regulatory proteins.

3.   a.   Generally, transcription factors bind to multiple sites in a cooperative manner and their effect on transcription is synergistic. The *GAL1* UAS element has four binding sites for the Gal4 protein. The UAS elements near the *GAL2* and *GAL7* genes each have two Gal4 binding sites. Loss of one Gal4 binding site from the *GAL1* UAS element may have a small effect on *GAL1* transcription, but you would still expect to see Gal4-specific transcriptional activation due to the three remaining binding sites.

      b.   The deletion of all four Gal4-binding sites from the UAS would prevent transcriptional activation of the *GAL1* gene. Experiments have shown that Gal4-DNA binding at the UAS is necessary for its function.

      c.   The binding of Mig1 at its binding site is necessary for the repression of *GAL1* in the presence of glucose. In this case, the absence of the Mig1 binding site will have no effect on *GAL1* transcription in the presence of galactose and the absence of glucose. It would effect *GAL1* transcription if both sugars were present as transcription would still be activated.

      d.   If the Gal4 protein is missing its activation domain, it would still bind at the UAS but it would not promote transcription. The activation domain helps recruit the transcriptional machinery to the promoter.

      e.   The product of the *GAL80* gene inhibits Gal4 function by binding to a region within the Gal4 activation domain. Therefore, the deletion of *GAL80* will have no effect on the transcription of *GAL1* in the presence of galactose.

      f.   The deletion of the *GAL1* promoter would prevent transcription of the gene. The promoter is required for the binding of the TATA-binding protein and then by extension, the binding of RNA polymerase II.

      g.   The product of the *GAL3* gene is both a sensor and inducer. It binds galactose (and ATP) causing an allosteric change that allows Gal3 to bind Gal80, which in turn releases Gal4 from Gal80. When Gal80 is bound to Gal4, Gal4 cannot activate transcription so its release by Gal80 is necessary for gene activation. In the absence of Gal3 protein, Gal4 will remain bound to Gal80 and will not activate *GAL1* transcription.

4.   In the presence of both galactose and glucose, the activation of the *GAL1* gene is prevented by the Mig1 protein. Mig1 is a sequence-specific DNA-binding protein that binds to a site between the UAS element and the promoter of the *GAL1* gene. Mig1 recruits a protein complex called Tup1 that contains a histone deacetylase and represses gene transcription.

5.   The acetylation of histones is one of several types of covalent modifications that are part of what is called a histone code. Histones associated with nucleosomes of active genes are hyperacetylated, whereas they are hypoacetylated in

nucleosomes associated with inactive genes. Given this, histone deacetylation plays a key role in gene repression while histone acetylation plays a key role in gene activation.

6. Among the mutations that would prevent an α strain of yeast from switching mating type would be mutations in the *HO* and *HMRa* genes. The *HO* gene encodes an endonuclease that cuts the DNA to initiate switching, and the *HMRa* locus contains the "cassette" of unexpressed genetic information for the MATa mating type.

7. In a haploid α cell, the α1 protein, in a complex with MCM1 protein, binds to specific DNA sites and activates the transcription of α-specific genes. The α2 protein, in a complex with MCM1 protein, binds to specific DNA sites and represses a-specific genes.

8. In yeast, the Sir proteins (silent information regulators) are necessary to keep the genetic information in the *HMR* and *HML* cassettes silent. Their functions facilitate the condensation of chromatin and help lock up *HMR* and *HML* in chromatin domains that are inaccessible to transcriptional activators. Mutations in SIR genes are sterile since both *a* and α information is expressed in the absence of this specific gene silencing.

9. The term epigenetic inheritance is used to describe heritable alterations in which the DNA sequence itself is not changed. It can be defined operationally as the inheritance of chromatin states from one cell generation to next. Genomic imprinting, X-chromosome inactivation, and position-effect variegation are several such examples.

10. An enhancersome is a large protein complex that acts synergistically to activate transcription. Because this is a case where the whole is greater than the sum of its parts, loss of any one of the required proteins may severely impact the required synergy.

11. Imprinted genes are functionally hemizygous. Maternally imprinted genes are inactive when inherited from the mother, and paternally imprinted genes are inactive when inherited from the father. A mutation in one of these genes is dominant when an offspring inherits a mutant allele from one parent and a "normal" but inactivated allele from the other parent.

12. A gene not expressed due to alteration of its DNA sequence will never be expressed and will be inherited generation to generation. An epigenetically inactivated gene may still be regulated. Chromatin structure can change in the course of the cell cycle, for example, when transcription factors modify the histone code. Also, inactivation may change generation to generation as it does for imprinted genes.

13. The inheritance of chromatin structure is thought to be responsible for the inheritance of epigenetic information. This is due to the inheritance of the histone code and may also include inheritance of DNA methylation patterns.

14. Many DNA-protein interactions are shared by prokaryotes and eukaryotes, but the mechanisms by which proteins bound to DNA at great distances from the start of transcription affect that transcription is unique to eukaryotes. Also, mechanisms of gene regulation based on chromatin structure are distinctly eukaryotic.

15. Transcription factors can be broadly divided into two classes: recruitment and activation of the transcriptional apparatus, and chromatin remodeling or modifying enzymes. Importantly, many of the transcriptional complexes that form share many subunits and appear to be assembled in a modular fashion. The many different combinatorial interactions that are possible can result in distinct patterns of gene expression.

16. a. D through J — the primary transcript will include all exons and introns.

    b. E, G, I — all introns will be removed.

    c. A, C, L — the promoter and enhancer regions will bind various transcription factors that may interact with RNA polymerase.

CHALLENGING PROBLEMS

17. There are numerous possibilities. As one example:

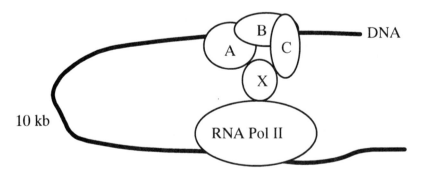

18. The following represents a cell that is mutant for TFC. In this case, CRX does not bind and the activation of transcription does not occur.

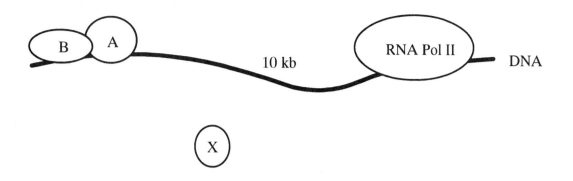

19. The same as 18 but now the binding site for TFC is mutant.

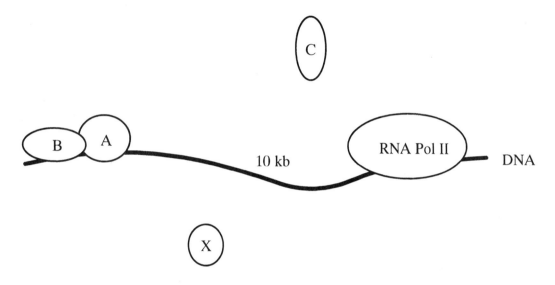

20. A gene not expressed due to alteration of its DNA sequence will never be expressed and will be inherited generation to generation. An epigenetically inactivated gene may still be regulated. Chromatin structure can change in the course of the cell cycle, for example, when transcription factors modify the histone code.

21. Epigenetic marks include both DNA methylation and histone modifications. Heritable epigenetic marks on histones and DNA have profound effects on chromatin structure and gene expression. Heterochromatin is associated with nucleosomes containing methylated histone H3. Proteins, such as HP-1 (heterochromatin protein-1) interact with specific epigenetic marks and are required in some way to produce or maintain heterochromatin. Other proteins act in concert with specific DNA sequences to insulate regions of euchromatin from regions of heterochromatin. Since epigenetic marks are inherited, chromatin structure is also inherited. Marked histones from nucleosomes of the parental strands are mixed with new, unmarked histones so that both daughter strands become associated with nucleosomes that contain both marked and unmarked histones. The code carried by the old histones most likely guides the modification of the new histones, ensuring that the chromatin state is remembered and passed on in future divisions.

22. Chromatin structure has a profound effect on gene expression. Transgenes inserted into regions of euchromatin would more likely be capable of expression than those inserted into regions of heterochromatin.

23. If there is an operon governing both genes, then a frameshift mutation could cause the stop codon separating the two genes to be read as a sense codon. Therefore, the second gene product will be incorrect for almost all amino acids. However, there are no known polycistronic messages in eukaryotes. The alternative, and better, explanation is that both enzymatic functions are performed by the same gene product. Here, a frameshift mutation beyond the first function, carbamyl phosphate synthetase, will result in the second half of the protein molecule being nonfunctional.

24. a. Clone A is activated by FGF, but B, C, and D are not. This indicates that the DNA binding site for this activation is located somewhere between the 3′ end of exon 1 [E1(f)] and the 5′ end of exon 3 [E3(f)]. This is the region of DNA found only in clone A.

    b. Cortisol represses transcription of clone A and B, but not C. (You do not expect any effect on D, the intact globin gene.) Comparing these clones indicates that the DNA site involved in this repression must be located in the 3′ region of E3(f) or the 3′ flanking sequences of this gene.

    c. Activation by EP is seen in clones C and D, but not B. (Again, you do not expect A, the intact c-fos gene to respond to EP.) This indicates that the DNA site involved in this activation must be localized to the 3′ side of E3(g) or the 3′ flanking regions of the globin gene.

12

The Genetic Control of Development

BASIC PROBLEMS

1. These are names of genes that are required for normal *Drosophila* development. The mutant phenotypes associated with these genes provide clues to their different roles: *knirps* is a gap gene; *runt* is a pair-rule gene; *gooseberry* is a segment polarity gene; and *antennapedia* is a segment identity (homeotic) gene.

2. The primary pair-rule gene *eve* (*even-skipped*) would be expressed in seven stripes along the A-P axis of the late blastoderm.

3. Pair-rule genes encode transcription factors expressed in repeating patterns of seven stripes. They are necessary to divide the embryo into the correct number of segments. Homeotic genes are also transcription factors, but they are essential for segment identity and do not affect segment number.

4. Proper *ftz* expression requires *Kr* in the fourth and fifth segments, and *kni* in the fifth and sixth segments. Each of the seven stripes of pair-rule gene expression is controlled independently through distinct cis-acting regulatory elements and unique combinations of trans-acting transcription factors.

5. A number of experiments could be devised. A comparison of amino acid sequence between mammalian gene products and insect gene products would indicate which genes are most similar to each other. Using cloned cDNA sequences from mammalian genes for hybridization to insect DNA would also indicate which genes are most similar to each other. The comparison of the gene and protein expression along the anterior-posterior axis in the two species may provide information relative to their deeply conserved functions. Finally, transgenic experiments could be performed to test functional equivalency. For example, when the mouse *HoxB6* gene is inserted in *Drosophila*, it can substitute for *Antennapedia*.

6. If you diagram these results, you will see that deletion of a gene that functions posteriorly allows the next-most anterior segments to extend in a posterior

direction. Deletion of an anterior gene does not allow extension of the next-most posterior segment in an anterior direction. The gap genes activate *Ubx* in both thoracic and abdominal segments, whereas the *abd-A* and *Abd-B* genes are activated only in the middle and posterior abdominal segments. The functioning of the *abd-A* and *Abd-B* genes in those segments somehow prevents *Ubx* expression. However, if the *abd-A* and *Abd-B* genes are deleted, *Ubx* can be expressed in these regions.

7.   For zygotically required recessive mutations, crossing +/*m* × +/*m* parents will produce offspring in the expected Mendelian ratio of 3 wild type : 1 mutant. Only homozygous mutant offspring will be mutant. For genes that exhibit a maternal effect, it is the genotype of the mother that determines the phenotype of the offspring. So for the cross outlined above (+/*m* × +/*m*), none of the offspring will be phenotypically mutant. However, a cross *of m/m* females to +/+ males will produce all mutant offspring, while a cross of +/+ females to *m/m* males will produce only wild-type offspring.

8.   The nonrescuable maternal-effect lethal mutations may produce a product that is required very early in development, before the developing fly is producing any proteins, while the rescuable maternal-effect lethal mutations may act later in development when embryo protein production can compensate for the maternal mutation.

9.   a.   A pair-rule gene.

      b.   Look for expression of the mRNA from the candidate gene in repeating pattern of seven stripes along the A-P axis of developing embryo.

      c.   No. An embryo mutant for the gap gene *Krüppel* would be missing many anterior segments. This effect would be epistatic to the expression of a pair-rule gene.

10.   Biocoid mRNA is tethered to the anterior pole of the egg. Upon translation, BCD protein diffuses in the common syncytial cytoplasm creating the observed gradient. As a transcription factor, the gradient of BCD protein provides positional information along the anteroposterior axis.

11.   All of the classes listed would show abnormal gene expression. A homozygous *bcd* mutant female would produce embryos lacking functional BCD protein. Functional BCD protein is necessary for correct regulation and expression of gap genes. Proper expression of gap genes is necessary for proper expression of pair-rule genes and homeotic genes, and the proper expression of pair-rule genes is necessary for the proper expression of segment-polarity genes.

CHALLENGING PROBLEMS

12. a. The homeodomain is a conserved protein domain containing 60 amino acids found in a significant number of transcription factors. Any protein that contains a functional homeodomain is almost certainly a sequence-specific DNA-binding transcription factor.

b. The *eyeless* gene (named for its mutant phenotype) regulates eye development in *Drosophila*. You would expect that it is expressed only in those cells that will give rise to the eyes. To test this prediction, visualization of the location of *eyeless* mRNA expression by *in situ* hybridization and the eyeless protein by immunological methods should be performed. Through genetic manipulation, it is possible to express the *eyeless* gene in tissues where it is not ordinarily expressed. For example, when *eyeless* is turned on in cells destined to form legs, eyes form on the legs!

c. Transgenic experiments have shown that the mouse *Small eye* gene and the *Drosophila eyeless* gene are so similar that the mouse gene can substitute for *eyeless* when introduced into *Drosophila*. As above, when the mouse *Small eye* gene is expressed in *Drosophila*, even in cells destined to form legs, eyes form on the legs! (Of course, the "eyes" are not mouse eyes as these genes act as a master switch to turn on the entire cascade of genes needed to build the eye, and in this case, the *Drosophila* set to build a *Drosophila* eye.)

13. a. The mutations are in distinct cis-acting regulatory elements of gene X. The A element controls expression of gene X in the brain, the B element controls heart expression, and the C element controls lung expression.

b.

gene *X*

c. Transgenic experiments using the three cis-acting regulatory elements hooked up to a reporter gene (for example, GFP, green fluorescent protein from jellyfish) will allow visualization of localized expression. In this case, you would expect a transgenic fly with the element A-GFP construct to show GFP expression in the brain. Other constructs should confirm the other expression patterns.

14. *Sonic hedgehog* (*Shh*) is expressed in several tissues of developing vertebrates, for example, the ZPA (zone of polarizing activity) of developing limb buds, the floor plate of the neural tube, and developing feather buds in chickens. These serve to illustrate the different roles played by toolkit genes, like *Shh*, at different places and times in development. Since these genes have multiple roles in different tissues and cell types, mutations in coding regions will likely have widespread defects. Mutations in regulatory regions will be dominant and viable because the wild-type allele of *Shh* will still supply the necessary functions while the mixexpressed mutant allele will cause the unexpected phenotype.

15. In *Drosophila*, the *dsx* transcript is alternatively spliced in the two sexes. The product of the *tra* gene is required for the female-specific RNA splicing of *dsx*. In the presence of Tra (and a related protein Tra2), the pre-mRNA of *dsx* is spliced to create the dsx^F transcript. In its absence, the dsx^M transcript is the default product of splicing. If a mutation prevents Tra from binding the *dsx* transcript, the default male-specific splicing will occur. In males, this will have no effect, but XX females homozygous for this mutation will be transformed into the male phenotype.

16. GLP-1 protein is localized to the two anterior cells of the four-cell *C. elegans* embryo by repressing its translation in the two posterior cells. The repression of GLP-1 translation requires the 3′ UTR spatial control region (SCR). Deletion of the SCR will allow *glp*-1 expression in both anterior and posterior cells. In both heterozygous and homozygous mutants, you would expect GLP-1 protein expression in all cells.

13

Genomes and Genomics

BASIC PROBLEMS

1. Contig describes a set of adjacent DNA sequences or clones assembled using overlapping sequences or restriction fragments. Because the pieces are assembled into a continuous whole, it makes sense that contig is derived from the word contiguous.

2. Because bacteria have relatively small genomes (roughly three megabase pairs) and essentially no repeating sequences, the whole genome shotgun approach would be used.

3. Terminal sequence reads of cloned inserts. such as those generated during whole genome shotgun sequencing, are assembled into a scaffold by matching homologous sequences shared by reads from overlapping clones. In essence, the central sequence of any single clone will be generated from the terminal sequences of overlapping clones.

4. A scaffold is also called a supercontig. Contigs are sequences of overlapping reads assembled into units, and a scaffold is a collection of joined-together contigs.

5. Yes. If the gap is short, PCR fragments can be generated from primers based on the ends of your contigs, and these fragments can be directly sequenced without a cloning step. The process does not work if the gap is too long.

6. The clone may contain DNA that hybidizes to a small family of repetitive DNA that is adjacent to the gene being studied or within one of its introns. Alternatively, the clone may share enough homology with members of a small gene family to hybridize to all or there may be four pseudogenes that are descendants of the unique gene being studied.

7. Yes. The clone hybridizes to and spans a translocation breakpoint that involves the X chromosome and an autosome. Because the DMD gene normally maps to

the X chromosome, it is of interest that a translocation of the X is found in a patient with DMD. If the translocation is the cause of DMD mutation, the clone identifies a least a portion and/or location of the gene.

8. You still need to verify that the gene with the amino acid change is the correct candidate for the phenotype under study. For example, you might transform mutants with a wild-type copy of the candidate gene and see if the transgene reverts the mutant phenotype (functional complementation). Alternatively, you can attempt to compare the phenotype of the mutant with the known function of the candidate gene. A third alternative solution is to determine if the presence of the amino acid change (or homozygosity for the amino acid change if the phenotype is recessive) correlates with the phenotype, and whether the change is absent in the general population that do not have the phenotype (or transmit the phenotype—if recessive).

9. Yes. The operator is the location at which repressor functionally binds through interactions between the DNA sequence and the repressor protein.

10. There are numerous possible drawings that can be made. The goal is to indicate that at least one exon be present in all eight genomic fragments and that the ESTs define the 5′ and 3′ ends of the transcript.

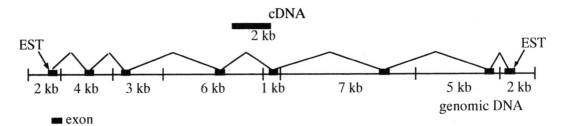

11. A level of 35 percent or more amino acid identity at comparable positions in two polypeptides is indicative of a common three-dimensional structure. It also suggests that the two polypeptides are likely to have at least some aspect of their function in common. However, it does not prove that your particular sequence does in fact encode a kinase.

12. The yeast two-hybrid test detects possible physical interactions between two proteins. These results indicate that gene A codes for a protein that interacts with proteins encoded by clones M and N, and further, that clone M encodes a protein that also interacts with proteins encoded by clones S and Q. For example:

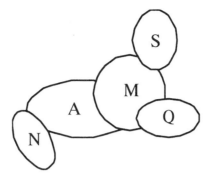

13. This is just a matter of aligning the sequences to determine their overlap.

Read 1:   TGGCCGTGATGGGCAGTTCCGGTG
Read 2:   TTCCGGTGCCGGAAAGA
Read 3: CTATCCGGGCGAACTTTTGGCCG
Read 4:   CGTGATGGGCAGTTCCGGTG
Read 5:   TTGGCCGTGATGGGCAGTT
Read 6:   CGAACTTTTGGCCGTGATGGGCAGTTCC

And this creates the contig:

CTATCCGGGCGAACTTTTGGCCGTGATGGGCAGTTCCGGTGCCGGAAAGA

14. You can determine whether the cDNA clone was a monster or not, by alignment of the cDNA sequence against the genomic sequence. (There are computer programs available to do this.) Is it derived from two different sites? Does the cDNA map within one [gene-sized] region in the genome or to two different regions? Of course, introns may complicate the issue.

15. a. Because the triplet code is redundant, changes in the DNA nucleotide sequence, (especially at those nucleotides coding for the third position of a codon) can occur without change to its encoded protein.

b. Protein sequences are expected to evolve and diverge more slowly than the genes that encode them. Due to the flexibility in the third positions of most codons, the DNA sequence can accumulate changes without affecting protein structure. Protein sequences will evolve and diverge more slowly because amino acid changes may change the structure and function of the protein. Since the amino acid sequence is functionally constrained, natural selection will eliminate many deleterious amino acid changes. This will reduce the rate of change in the amino acid sequence and lead to greater sequence conservation of the amino acid sequence compared to the nucleotide sequence.

16. Gene position is not a necessary requirement for using RNAi. What is needed is the sequence of the gene that you intend to inactivate, as you need to start with double-stranded RNA that is made with sequences that are homologous to part of the gene.

17. Phenocopying is the mimicking of a mutant phenotype by inactivating the gene product rather than the gene itself. It can be used regardless of how well-developed the genetic technology has been elaborated for a particular organism and can provide meaningful results when gene knockout or site-directed mutagenesis is not possible. Three methods of producing phenocopies are: the prevention of gene-specific translation by the introduction of anti-sense RNA (RNA complementary to the gene-specific mRNA); the introduction of double-stranded RNA with sequences homologous to part of a specific mRNA resulting in major reduction in the level of the mRNA (called double-stranded RNA interference); and inhibition of a specific protein through high-affinity binding of compound(s) identified through chemical genetics.

18. Forward genetics identifies heritable differences by their phenotypes and map locations and precedes the molecular analysis of the gene products. Reverse genetics starts with an identified protein or RNA and works toward mutating the gene that encodes it (and in the process, discovered the phenotype when the gene is mutated).

CHALLENGING PROBLEMS

19. **a.** This is just a matter of aligning the sequences to determine their overlap.

```
Read 1:                                             ATGCGATCTGTGAGCCGAGTCTTTA
Read 2:                      AACAAAAATGTTGTTATTTTTATTTCAGATG
Read 3:                              TTCAGATGCGATCTGTGAGCCGAG
Read 4:  TGTCTGCCATTCTTAAAAACAAAAATGT
Read 5:                          TGTTATTTTTATTTCAGATGCGA
Read 6:                      AACAAAAATGTTGTTATT
```

And this creates the contig

```
5´-TGTCTGCCATTCTTAAAAACAAAAATGTTGTTATTTTTATTTCAGATGCGATCTGTGAGCCGAGTCTTTA-3´
```

Since you do not know if the DNA sequence represents the template strand for the mRNA or is the complementary strand, there are two possible transcripts.

```
5´-UGUCUGCCAUUCUUAAAAACAAAAAUGUUGUUAUUUUUAUUUCAGAUGCGAUCUGUGAGCCGAGUCUUUA-3´
```

and

```
3´-ACAGACGGUAAGAAUUUUUGUUUUUACAACAAUAAAAAUAAAGUCUACGCUAGACACUCGGCUCAGAAAU-5´
```

b. Translation of the first possible transcript starting at the first letter reads

CLPFLKTKMLLFLFQMRSVSRVF

Translation starting at the second letter reads

```
VCHS stop
```

Translation starting at the third reads

```
SAILKNKNVVIFISDAICRPSL
```

For the second possible transcript, starting at the first letter reads

```
stop
```

For the second possible transcript, starting at the second letter reads

```
KDSAHRSHLK stop
```

For the second possible transcript, starting at the third letter reads

```
KTRLTDRI stop
```

    c.   Using the nucleotide sequence of the contig and performing BLASTn or the possible translation products and performing tBLASTn, you will discover that this sequence and the translation product above listed first match perfectly with a region of exon 19 of the human CFTR gene.

20. The correct assembly of large and nearly identical regions is problematic with either method of genomic sequencing. However, the whole genome shotgun method is less effective at finding these regions than the clone-based strategy. This method also has the added advantage of easy access to the suspect clone(s) for further analysis. (Comparative genomic hybridization has been used effectively and is superior to sequencing for this determination. See Chapter 16, figure 16-36 in the companion text.)

21. Assessing whether a short sequence constitutes an exon is difficult. The best way to determine if a suspected micro-exon is actually used is to look for a cDNA or an EST that includes it. Alternatively, identification of consensus donor and acceptor splice site sequences can be tried and also the use of comparative genomics, that is, the conservation of the predicted amino acid encoded by the micro-exon in the same or other genomes.

22. a. Forward genetics identifies heritable differences by their phenotypes and map locations and precedes the molecular analysis of the gene products. Reverse genetics starts with an identified protein or RNA and works toward mutating the gene that encodes it (and in the process, discovers the phenotype when the gene is mutated). Because you have known proteins and want to determine the phenotypes associated with loss-of-function mutations in the genes that encoded them, a reverse genetic approach is the answer.

b. The two general approaches would be either directed mutations in the gene of interest or the generation of phenocopies of the mutant phenotype by inactivating the gene product rather than the gene itself.

23. Fifteen percent are essential gene functions (such as enzymes required for DNA replication or protein synthesis).

Twenty-five percent are auxotrophs (enzymes required for the synthesis of amino acids or the metabolism of sugars, genes required for normal growth rate and normal colony morphology or other traits evident in the laboratory, etc.).

Sixty percent are redundant or pathways not tested (genes for histones, tubulin, ribosomal RNAs, etc., are present in multiple copies; the yeast may require many genes under only unique or special situations or in other ways that are not necessary for life in the "lab"—where the yeast will be exposed to competition from other microbes and a wide array of conditions not found in the laboratory).

24. **a.** The two *E. coli* strains K-12 and O157:H7 have 3574 protein-coding genes in common and among these orthologous genes, share 98.4 percent nucleotide identity. Even so, the complete genomes and proteomes of these species differ enormously. The genome of O157:H7 contains 1387 genes not found in the K-12 genome, and there are 528 genes in the K-12 genome not found in the O157:H7 genome. Most of the O157:H7-specific sequences are horizontally transferred foreign DNAs and would be expected to contain candidate genes that encode virulence factors, cell-invasion proteins, adherence proteins, secretion proteins, and possible metabolic genes required for nutrient transport, antibiotic resistance, and other activities that may confer the ability to survive in different hosts. It would certainly be expected that comparisons between K-12 and uropathogenic strains of *E. coli* would show similar enormous differences in their genomes and proteomes.

b. In comparing O157:H7 to uropathogenic strains of *E. coli*, you would expect that they would share the same backbone genes as K-12 and also possibly share some of the candidate genes that encode virulence factors, cell-invasion proteins, adherence proteins, secretion proteins, and other activities that help confer the ability to survive in different hosts. However, you would also expect significant differences between the two genomes as the urinary tract and colon represent separate niches that likely require unique gene products.

c. The various strains of *E. coli* have evolved through a complex process. The ancestral backbone genes that define *E. coli* have undergone slow accumulation of vertically acquired sequence changes. This can be seen in the 98.4 percent nucleotide identity in the orthologous genes shared by K-12 and O157:H7. However, the sequences not in common have been introduced via numerous, independent horizontal gene-transfer events at many discrete sites. These sequences are from the genomes of viruses and other bacterial species. Some differences between the strains may also be the result of deletion.

d.   Functions of genes may be determined by various mechanisms. Comparison of the strain-specific genes to all other known genes using a BLAST search might provide insight to function. Reverse genetic analysis to disrupt gene function and observation of the resulting phenotype may allow you to assess the role of the normal gene product. Finally, transforming E. *coli* K-12 with pathogenic-specific genes and analysis of the resulting transgenic strains may also allow gene function to be assessed.

14

The Dynamic Genome

1. R plasmids are the main carriers of drug resistance. These plasmids are self-replicating and contain any number of genes for drug resistance, as well as the genes necessary for transfer by conjugation (called the RTF region). It is R plasmid's ability to transfer rapidly to other cells, even those of related species, that allows drug resistance to spread so rapidly. R plasmids acquire drug-resistance genes through transposition. Drug-resistance genes are found flanked by IR (inverted repeat) sequences and as a unit are known as transposons. Many transposons have been identified, and as a set they encode a wide range of drug resistances (see the table in the companion text). Because transposons can "jump" between DNA molecules (e.g., from one plasmid to another or from a plasmid to the bacterial chromosome and vice versa), plasmids can continue to gain new drug-resistance genes as they mix and spread through different strains of cells. It is a classic example of evolution through natural selection. Those cells harboring R plasmids with multiple drug resistances survive to reproduce in the new environment of antibiotic use.

2. Boeke, Fink, and their co-workers demonstrated that transposition of the Ty element in yeast involved an RNA intermediate. They constructed a plasmid using a Ty element that had a promoter that could be activated by galactose, and an intron inserted into its coding region. First, the frequency of transposition was greatly increased by the addition of galactose, indicating that an increase in transcription (and production of RNA) was correlated to rates of transposition. More importantly, after transposition they found that the newly transposed Ty DNA lacked the intron sequence. Because intron splicing occurs only during RNA processing, there must have been an RNA intermediate in the transposition event.

3. P elements are transposable elements found in *Drosophila*. Under certain conditions they are highly mobile and can be used to generate new mutations by random insertion and gene knockout. As such, they are a valuable tool to tag and then clone any number of genes. P elements can also be manipulated and used to insert almost any DNA (or gene) into the *Drosophila* genome. P element-

mediated gene transfer requires inserting the DNA of interest between the inverted repeats necessary for P element transposition. This recombinant DNA, along with helper intact P element DNA (to supply the transposase), are then co-injected into very early embryos. The progeny of these embryos are then screened for those that contain the randomly inserted DNA of interest.

4. Most eukaryotic class 2 elements transpose by a "cut and paste" mechanism that involves their excision from one site and reinsertion into a new site elsewhere in the genome. Class 2 elements can increase their copy number in a few ways; one is by excising from a site that has been replicated into another site that has not yet been replicated. On the other hand, for class 1 elements, the RNA transcript is the transposition intermediate (it is copied into double-stranded DNA and reinserted into the genome). As such, a single class 1 element can give rise to many transcripts, and each can theoretically be copied into double stranded DNA and reinserted. However, because the reinsertion of so many DNA copies into the genome would almost certainly harm the host, mechanisms have evolved to prevent this from happening. For example, the transcription of Class 1 elements is usually repressed. In addition, in situations where transcription of Class 1 elements is not repressed, the host controls the conversion of the RNA transcript into double-stranded DNA.

5. P elements are transposable elements found in *Drosophila*. Under certain conditions they are highly mobile and can be used to generate new mutations by random insertion and gene knockout. As such, they are a valuable tool to tag and then clone any number of genes. P elements can also be manipulated and used to insert almost any DNA (or gene) into the *Drosophila* genome. P element mediated gene transfer requires inserting the DNA of interest between the inverted repeats necessary for P element transposition. This recombinant DNA, along with helper intact P element DNA (to supply the transposase), are then co-injected into very early embryos. The progeny of these embryos are then screened for those that contain the randomly inserted DNA of interet.

6. Some transposable elements have evolved strategies to insert into safe havens—regions of the genome where they will do minimal harm. Safe havens include duplicate genes (such as tRNA or rRNA genes) and other transposable elements. Safe havens in bacterial genomes might be very specific sequences between genes or the repeated rRNA genes.

7. Politics aside, Barbara McClintock was classically ahead of her time. She once wrote, "I stopped publishing detailed reports long ago when I realized, and acutely, the extent of disinterest and lack of confidence in the conclusions I was drawing from the studies." But beginning in the 1970's, transposition was discovered in bacteria and then in yeast. Transposable elements were subsequently found to be a significant component of most genomes, not just an oddity found only in maize. The importance of transposition in the transfer of drug resistance, cancer, and genetic engineering, just to name a few, was finally

understood at the molecular level, and she was fully recognized for her lifetime of brilliant work.

CHALLENGING PROBLEMS

8. **a.** The consequences will be different for different genes and different insertions. In the simplest scenario, the insertion prevents the binding of transcriptional activators that are required for the ultimate binding of RNA polymerase to the promoter. In this case, the gene will not be expressed (no mRNA will be synthesized).

   In more complex scenarios, the gene may be regulated by many enhancers (as is the case for most human genes). For example, one enhancer might be required for transcription in the liver, one required for transcription in muscle cells, etc. In this instance, insertion of the LINE into the liver enhancer may prevent transcription of the gene in the liver but not interfere with its transcription in muscle.

   b. Again, the consequences will differ depending on the gene that has sustained the insertion. In the simplest scenario, the presence of the transposable element will provide a binding site for the transcriptional repressor to bind near the promoter and prevent the binding of RNA polymerase II.

   c. The *Alu* element will be transcribed into RNA with the rest of the gene sequences and will prevent the splicing of the intron that it has inserted into. The insertion will almost certainly result in a null allele as the *Alu* sequence and the intron will now be translated. The intron, *Alu,* or both probably will contain stop codons.

   d. The insertion and excision will result in a 3-bp indel in the exon and slightly alter the amino acid sequence of the protein, but it will not produce a frameshift mutation. The minor change in the amino acid sequence may or may not affect the function of the encoded protein.

   e. This insertion and excision will produce a frameshift mutation and is more likely than (d) to impair protein function.

   f. Chances are that this mutation will not affect gene expression as the intron will probably still be spliced correctly.

9. The staggered cut will lead to a nine base-pair target site duplication that flanks the inserted transposon.

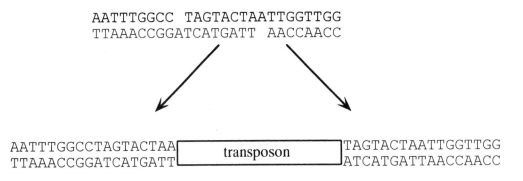

```
AATTTGGCC TAGTACTAATTGGTTGG
TTAAACCGGATCATGATT AACCAACC
```

```
AATTTGGCCTAGTACTAA                    TAGTACTAATTGGTTGG
TTAAACCGGATCATGATT  | transposon |    ATCATGATTAACCAACC
```

10. It would not be surprising to find a SINE element in an intron of a gene, rather than an exon. Processing of the pre-mRNA would remove the transposable element as part of the intron, and translation of the FB enzyme would not be effected.

11. In the *Ac-Ds* system, *Ac* can produce an unstable allele that is autonomous. *Ds* can revert only in the presence of *Ac* and is nonautonomous. In the following, *Ac+* indicates the absence of the *Ac* regulator gene.

Cross 1:

P C/c^{Ds}; Ac/Ac^+ × c/c; Ac^+/Ac^+

F$_1$ $1/4$ C/c; Ac/Ac^+ (solid pigment)

          $1/4$ C/c; Ac^+/Ac^+ (solid pigment)

          $1/4$ c^{Ds}/c; Ac/Ac^+ (unstable colorless or spotted)

          $1/4$ c^{Ds}/c; Ac^+/Ac^+ (colorless)

Overall: 2 solid:1 spotted:1 colorless

Cross 2:

P C/c^{Ac} × c/c

F$_1$ $1/2$ C/c (solid pigment)

          $1/2$ C/c^{Ac} (unstable colorless or spotted)

Overall: 1 solid:1 spotted

Cross 3:

P C/c^{Ds}; Ac/Ac^+ × C/c^{Ac}; Ac^+/Ac^+

F$_1$ $1/8$ C/C; Ac/Ac^+ (solid pigment)

          $1/8$ C/c^{Ac}; Ac/Ac^+ (solid pigment)

          $1/8$ C/C; Ac^+/Ac^+ (solid pigment)

$1/8\ C/c^{Ac};Ac^+/Ac^+$ (solid pigment)

$1/8\ C/c^{Ds};Ac^+/Ac^+$ (solid pigment)

$1/8\ C/c^{Ds};Ac^+/Ac$ (solid pigment)

$1/8\ c^{Ds}/c^{Ac};Ac^+/Ac^+$ (unstable colorless or spotted)

$1/8\ c^{Ds}/c^{Ac};Ac^+/Ac$ (unstable colorless or spotted)

Overall: 3 solid:1 spotted

12. It would not be surprising to find a SINE element in an intron of a gene, rather than an exon. Processing of tbe pre-mRNA would remove the transposable element as part of the intron, and translation of the FB enzyme would not be effected.

13. There is no correct answer to this question. Yeast does have a very compact genome with almost 70 percent of its DNA representing exons, so it may not contain sufficient "safe havens" for DNA transposition. Transposons (and transposition events) that kill the host organism do not have a chance to spread. The *Ty* elements of yeast have evolved mechanisms that target their insertions so that they presumably do not harm their host. Perhaps it has only been chance that DNA transposons are not found in yeast but it is also possible that yeast has specific mechanisms that prevent their spread. One could look at related species of yeasts to determine if they have class 2 elements.

15

Mutation, Repair, and Recombination

BASIC PROBLEMS

1. You need to know the reading frame of the possible message.

2. The mutant has a deletion of one base and this will result in a frameshift (−1) mutation.

3. Proline can be coded for by CCN (N stands for any nucleotide) and histidine can be coded for by CAU or CAC. For a mutation to change a proline codon to a histidine codon requires a transverion (C to A) at the middle position. Therefore, a transition-causing mutagen cannot cause this change. Serine can be coded for by UCN. A change from C to U at the first position (a transition) would cause this missense mutation and would be possible with this mutagen.

4. Assuming single base-pair substitutions, then CGG can be changed to CGU, CGA, CGC, or AGG and still would code for arginine.

5. **a. and b.** By transversion, CGG (arginine) can be become AGG (arginine), GGG (glycine), CCG (proline), CUG (leucine), CGC (arginine), or CGU (arginine).

6. The enol form of thymine forms three hydrogen bonds with guanine.

7. **a.** If the enol form of thymine is inserted during replication. it will be inserted opposite a G. During the next round of replication, one expects the thymine to revert back to its keto form, and to specify the insertion of an A in the opposite strand. So the transition eill be from a C-G base pair (prior to insertion of the T) to a T-A base pair (after fixation of the mutation).

   b. After the first round of replication, the enol form of thymine will be paired with guanine. At the next round, the guanine will pair with cytosine to result in the TA•CG transiton.

8. **a.** Acridine orange causes frameshift mutations and frameshift mutations often result in null alleles.

   b. A +1 frameshift mutation can be reverted by two further single insertions so that the reading frame is re-established.

9. The following is a list of observations that argue "cancer is a genetic disease":

   1. Certain cancers are inherited as highly penetrant simple Mendelian traits.

   2. Most carcinogenic agents are also mutagenic.

   3. Various oncogenes have been isolated from tumor viruses.

   4. A number of genes that lead to the susceptibility to particular types of cancer have been mapped, isolated, and studied.

   5. Dominant oncogenes have been isolated from tumor cells.

   6. Certain cancers are highly correlated to specific chromosomal rearrangements. (See Chapter 16 of the companion text.)

10. XP (xeroderma pigmentosum) patients lack nucleotide excision repair and are highly prone to developing pigmented skin cancers. Individuals with HNPCC (hereditary nonpolyposis colorectal cancer) are prone to colorectal cancer due to a loss of the mismatch repair system. Individuals homozygous for mutations in *BRCA1* or *BRCA2* (breast cancer predisposition genes 1 and 2) are prone to breast cancer due to the loss of repair of double-stranded breaks.

11. DNA in *E. coli* is methylated. To distinguish the old, template strand from the newly synthesized strand, the mismatch repair mechanism takes advantage of a delay in the methylation of the new strand. This makes sense as replilcation errors produce mismatches only on the newly synthesized strand, so the mismatch repair system replaces the "wrong" base on that strand.

12. The mismatched "T" would be corrected to C and the resulting ACG, after transcription, would be 5´ UGC 3´ and code for cysteine. Or if the other strand was corrected, ATG would be transcribed to 5´ UAC 3´ and code for tyrosine.

13. NHEJ (nonhomologous end-joining) is error prone as some sequence may be lost in the repair process. The consequences of imperfect repair may be far less harmful than leaving the lesion unrepaired. Presumably this repair pathway evolved because, unless repaired, the broken ends can degrade further, leading to loss of more genetic information. Also, these lesions can initiate potentially harmful chromosomal rearrangements that could lead to cell death.

14. There is a very strong correlation between mutagens and carinogens. As discussed in Problem 9, cancer is a genetic disease. Therefore, any chemical classified as a mutagen by the Ames test should also be considered a carcinogen.

15. Meiotic recombination is initialized when the SpoII protein makes double-strand cuts in one of the homologous chromosomes. It is highly conserved in eukaryotes, indicating that this mechanism to initiate recombination is also conserved. Bacterial species do not have reciprocal meiotic recombination, so you would not expect this function to be conserved.

16. **a.** A transition mutation is the substitution of a purine for a purine or the substitution of a pyrimidine for a pyrimidine. A transversion mutation is the substitution of a purine for a pyrimidine, or vice versa.

    b. Both are base-pair substitutions. A synonymous mutation is one that does not alter the amino acid sequence of the protein product from the gene, because the new codon codes for the same amino acid as did the nonmutant codon. A neutral mutation results in a different amino acid that is functionally equivalent, and the mutation therefore has no known adaptive significance.

    c. A missense mutation results in a different amino acid in the protein product of the gene. A nonsense mutation causes premature termination of translation, resulting in a shortened protein.

    d. Frameshift mutations arise from addition or deletion of one or more bases in other than multiples of three, thus altering the reading frame for translation. Therefore, the amino acid sequence from the site of the mutation to the end of the protein product of the gene will be altered. Frameshift mutations can and often do result in premature stop codons in the new reading frame, leading to shortened protein products. A nonsense mutation causes premature termination of translation in the original reading frame, resulting in a shortened protein.

17. If an accidental double-strand break occurs after replication, it may be repaired by SDSA (synthesis dependent strand annealing). During this process, the RAD51-DNA filament "searches" the undamaged sister chromatid for the complementary sequence that will be used as the template for error-free repair. (If it occurs before synthesis, then the error-prone NHEJ repair mechanism might be used.) During meiosis, RAD51 associates with the meiosis-specific protein, Dmc1. The RAD51-Dmc1-DNA filament is now capable of "finding" its complementary sequence on the homologous chromosome, rather than its sister chromatid. Also, the "repair" of SpoII-generated double-strand breaks during meiosis may result in recombination.

18. Gene conversion is a type of nonreciprocal transfer of genetic information. If it occurs during meiosis, it alters the expected Mendelian ratios. During

recombination, small regions of heteroduplex DNA are formed (see Figure 15-34 in the companion text). For any given heterozygous gene, the heteroduplex formed may occasionally include the sequence differences associated with the two alleles. If so, these sequence differences will create mismatched base pairs in the newly formed heteroduplex DNA and will represent a target for mismatch repair. Mismatch repair can either "correct" the invading strand so that the original sequence (and allele) is maintained, or equally likely, repair the original strand (using the invading strand as template) and, in the process, "convert" the allele.

19. Depurination results in the loss of the adenine or guanine base from the DNA backbone. Because the resulting apurinic site cannot specify a complementary base, replication is blocked. Under certain conditions, replication proceeds with a near random insertion of a base opposite the apurinic site. In three-fourths of these insertions, a mutation will result.

   Deamination of cytosine yields uracil. If left unrepaired, the uracil will be paired with adenine during replication, ultimately resulting in a transition mutation.

   Deamination of 5-methylcytosine yields thymine and thus frequently leads to C to T transitions.

   Oxidatively damaged bases, such as 8-OxodG (8-oxo-7-hydrodeoxyguanosine) can pair with adenine, resulting in a transversion.

   Errors during DNA replication can lead to spontaneous indel mutations.

20. Translesion or bypass polymerases are able to replicate past damaged DNA that otherwise would stall replicative polymerases. They differ from replicative polymerases in that they can tolerate large adducts on the bases (as they have much larger active sites that can accommodate damaged bases), they are much more error-prone (as they lack the 3′ to 5′ proofreading function), and they can only add relatively few nucleotides before falling off. Their main function is to unblock the replication fork, not to synthesize long stretches of DNA that could contain many mismatches.

21. During recombination, small regions of heteroduplex DNA are formed when a strand from one homologous chromosome "invades" the other (see figure 15-34 in the companion text). Since is unlikely that the two homologous chromosomes are identical, the heteroduplex formed by the base-pairing of one strand from one homolog with the complementary strand from the other homolog, may contain sequence mismatches. For any given heterozygous gene, the heteroduplex formed may occasionally include the sequence differences associated with the two alleles. If so, these sequence differences will create mismatched base pairs in the newly formed heteroduplex DNA and will represent a target for mismatch repair. Mismatch repair can either "correct" the invading strand so that the original sequence (and allele) is maintained, or

equally likely, repair the original strand (using the invading strand as template), and in the process, "convert" the allele to the alternative form that is present in the invading strand.

22. There are many repair systems that are available: direct reversal, excision repair, transcription-coupled repair, and non-homologous end-joining.

23. Yes. It will cause CG-to-TA transitions.

24. Since cells of higher eukaryotes are usually not replicating their DNA, error-free repair is not possible because there are no undamaged strands or sister chromatids available as templates for new DNA synthesis.

25. DNA damage that stalls transcription is repaired by TC-NER (transcription-coupled nucleotide excision repair). Humans lacking this pathway suffer from Cockayne syndrome. A consequence of this defect is that a cell is much more likely to activate its apoptosis (cell suicide) pathway. Affected individuals are very sensitive to sunlight and have short stature, the appearance of premature aging, and a variety of developmental disorders.

CHALLENGING PROBLEMS

26. a. Because 5´-UAA-3´ does not contain G or C, a transition to a GC pair in the DNA cannot result in 5´-UAA-3´. 5´-UGA-3´ and 5´-UAG-3´ have the DNA antisense-strand sequence of 3´-ACT-5´ and 3´-ATC-5´, respectively. A transition to either of these stop codons occurs from the nonmutant 3´-ATT-5´, respectively. However, a DNA sequence of 3´-ATT-5´ results in an RNA sequence of 5´-UAA-3´, itself a stop codon.

    b. Yes. An example is 5´-UGG-3´, which codes for Trp, to 5´-UAG-3´.

    c. No. In the three stop codons the only base that can be acted upon is G (in UAG, for instance). Replacing the G with an A would result in 5´-UAA-3´, a stop codon.

27. a. and b. Mutant 1: most likely a deletion. It could be caused by radiation.

    Mutant 2: because proflavin causes either additions or deletions of bases and because spontaneous mutation can result in additions or deletions, the most probable cause was a frameshift mutation by an intercalating agent.

    Mutant 3: 5-BU causes transitions, which means that the original mutation was most likely a transition. Because HA causes GC-to-AT transitions and HA cannot revert it, the original must have been a GC-to-AT transition. It could have been caused by base analogs.

Mutant 4: the chemical agents cause transitions or frameshift mutations. Because there is spontaneous reversion only, the original mutation must have been a transversion. X-irradiation or oxidizing agents could have caused the original mutation.

Mutant 5: HA causes transitions from GC-to-AT, as does 5-BU. The original mutation was most likely an AT-to-GC transition, which could be caused by base analogs.

c. The suggestion is a second-site reversion linked to the original mutant by 20 map units and therefore most likely in a second gene. Note that auxotrophs equal half the recombinants.

28. a. A lack of revertants suggests either a deletion or an inversion within the gene.

b. To understand these data, recall that half the progeny should come from the wild-type parent.

Prototroph A: because 100 percent of the progeny are prototrophic, a reversion at the original mutant site may have occurred.

Prototroph B: half the progeny are parental prototrophs, and the remaining prototrophs, 28 percent, are the result of the new mutation. Notice that 28 percent is approximately equal to the 22 percent auxotrophs. The suggestion is that an unlinked suppressor mutation occurred, yielding independent assortment with the *nic* mutant.

Prototroph C: there are 496 "revertant" prototrophs (the other 500 are parental prototrophs) and four auxotrophs. This suggests that a suppressor mutation occurred in a site very close to the original mutation and was infrequently separated from the original mutation by recombination [$100\%(4 \times 2)/1000 = 0.8$ m.u.].

29. Compare the original amino acid sequences to the mutant ones and list the changes.

    original: ile;   mutant: met
    original: asn;   mutant: ser
    original: stop; mutant: trp

Now compare the codons that must have been altered by this mutagen.

    original: ile AU<u>A</u>;   mutant: met AU<u>G</u>
    original: asn A<u>A</u>C or A<u>A</u>U;   mutant: ser A<u>G</u>C or A<u>G</u>U
    original: stop U<u>A</u>G or UGA; mutant: trp U<u>G</u>G

All these mutations can be the result of a T to C transition in the DNA. This would result in the A to G change in the mRNA that explains all three codon changes. This mutagen, then, might work by altering the base-pairing specificity of T so that it now base pairs with G. Or, the mutagen could cause A to G transitions in the non-template strand, which would have the same effect.

30. Yes. Because DNA is a double-stranded molecule, replication of the DNA strand with a T to T* (altered T that base pairs with G) change produces an A to G transition in the newly replicated complementary DNA strand. If the mRNA is transcribed from the strand with the A to G change (the template strand), a U to C change is produced in the corresponding mRNA.

original: leu C<u>U</u>N;   mutant: pro C<u>C</u>N (where N = any base)

31. The new "wild-type" isolate contains an allele for a gene that increases the spontaneous reversion rate of *ad-3*. This allele appears to be unlinked to *ad-3* because independent assortment is observed. Call this gene *rev*. The cross becomes

ad-3 ; *rev*$^+$ (Strain A) × *ad-3*$^+$; *rev* (wild-type isolate)

Of the *ad-3* progeny, half would be *rev*$^+$ (low reversion rate) and half would be *rev* (high reversion rate).

Further crosses to map *rev* could be performed as well as testing the allele and gene specificity of the high reversion rate to see if it is the result of increased spontaneous mutation rates (gene non-specific) or some other process (*ad-3* specific).

32. XP is a heterogeneous genetic disorder and is caused by mutations in any one of several genes involved in the process of NER (nucleotide excision repair). As you read in the text about the discovery of yet another protein involved in NHEJ through research on cell line 2BN, it is certainly possible that this new patient has a mutation in as yet an unknown gene that encodes a protein necessary for NER.

33. Sun worshippers beware! Science has established a clear link between UV exposure and skin cancer, and with the drastic changes occurring in the protective ozone layer, UV levels are rising. UV damages DNA causing our cells to struggle to repair this damage by a number of different mechanisms. While this damage may be so great that the cell dies (the peeling of dead cells after asunburn), other times the damage is more insidious as it causes permanent changes in the instructions encoded in the DNA itself. These changes are sometimes caused by the cell attempting to repair the damage. In these cases, when the damage prevents the DNA from being "read," the cell may simply "guess" as to the meaning, changing the instructions. Eventually, the changed instructions may tell a cell to divide when it's not supposed to, or not die when it is supposed to. These are the very changes that make a cell cancerous.

34. We expect the linear asci to be 4:4 (4 *arg*:4 *arg*⁺). Gene conversion is indicated by departures from this ratio: 6:2 (=2:6), 5:3 (=3:5), and 3:1:1:3 (known as aberrant 4:4). Asci 3, 4, and 6 show gene conversion.

35. The first ascus shows conversion of all three mutant sites. During recombination, the region of heteroduplex DNA can be of some length, resulting in the co-conversion of tightly-linked markers. This must be the case here as all three mutant sites are showing the same 2:6 ratio. For the second ascus, only mutant site 3 shows conversion. In this case, the region of heteroduplex did not include mutant sites 1 and 2, or if it did, these mismatches were "repaired" back to the mutant sequences.

36. For the following, *A* and *a* will be used as the markers and for simplicity, only one possible starting combination will be shown.

Start, no repair	Repair of first heteroduplex		Repair of second heteroduplex		Repair of both heteroduplexes			
A	*A*	*A*	*A*	*A*	*A*	*A*	*A*	*A*
A	*A*	*A*	*A*	*A*	*A*	*A*	*A*	*A*
A	*A*	*a*	*A*	*A*	*A*	*A*	*a*	*a*
a	*A*	*a*	*a*	*a*	*A*	*A*	*a*	*a*
A	*A*	*A*	*A*	*a*	*A*	*a*	*A*	*a*
a	*a*	*a*	*A*	*a*	*A*	*a*	*A*	*a*
a	*a*	*a*	*a*	*a*	*a*	*a*	*a*	*a*
a	*a*	*a*	*a*	*a*	*a*	*a*	*a*	*a*

16

Large-Scale Chromosomal Changes

BASIC PROBLEMS

1. MM N OO would be classified as $2n-1$; MM NN OO would be classified as $2n$; and MMM NN PP would be classified as $2n+1$.

2. It would more likely be an autopolyploid. To make sure it was polyploid, you would need to microscopically examine stained chromosomes from mitotically dividing cells and count the chromosome number.

3. Aneuploid. Trisomic refers to three copies of one chromosome. Triploid refers to three copies of all chromosomes.

4. There would be one possible quadrivalent.

5. No. Amphidiploid means "doubled diploid." Because cauliflower has $n = 9$ chromosome, it could not have arose in this fashion. It has, however, contributed to other amphidiploid species.

6. The progenitor had nine chromosomes from a cabbage parent and nine chromosomes from a radish parent. These chromosomes were different enough that pairs did not synapse and segregate normally at meiosis. By doubling the chromosomes in the progenitor ($2n = 36$), all chromosomes now had homologous partners and meiosis could proceed normally.

7. In wheat, $2n = 6x = 42$ so B (or x) represents seven chromosomes.

8. Cross *T. tauschii* and Emmer to get ABD offspring. Treat the offspring with colchicine to double the chromosome number to AABBDD to get the hexaploid bread wheat.

9. Cells destined to become pollen grains can be induced by cold treatment to grow into embryoids. These embryoids can then be grown on agar to form monoploid plantlets.

10. The likely origin of a disomic is nondisjunction during meiosis. Depending whether the nondisjunction took place during the first or second division, you would expect one nullosomic, or two nullosomics and another disomic, respectively.

11. Yes. You would expect that one-sixth of the gametes would be *a*. Also, two-sixths would be *A*, two-sixths would be *Aa*, and one-sixth would be *AA*.

12. Both XXY and XXX would be fertile. XO (Turner) and XXY (Klinefelter) are both sterile.

13. Older mothers have an elevated risk of having a child with Down syndrome and other non-disjunctional events.

14. No. The DNA backbone has strict 5′ to 3′ polarity, and 5′ ends can only be joined to 3′ ends.

15. A crossover within a paracentric inversion heterozygote results in a dicentric bridge (and an acentric fragment).

16. By definition, an acentric fragment has no centromere, so it cannot be aligned or moved during meiosis (or mitosis). Consequently, at the end of a cell division, it gets left in the cytoplasm where it is not replicated.

17. possible translocation

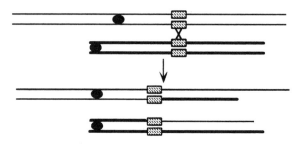

possible deletion (and duplication)

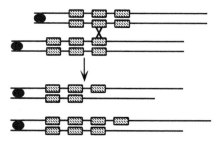

18. You could cross a strain with the appropriate 3; 4 reciprocal translocation to a wild-type strain to generate heterozygotes. One of the meiotic products of adjacent-1 segregation in these heterozygotes will have a duplication of the translocated portion of chromosome 3 and a deletion for the translocated portion of chromosome 4.

19. Very large deletions tend to be lethal. This is likely due to genomic imbalance or the unmasking of recessive lethal genes. Therefore, the observed very large pairing loop is more likely to be from a heterozygous inversion.

20. Because the new mutant does not show pseudodominance with any of the deletions that span chromosome 2, it is likely that the mutation does not map to this chromosome or it is located in a region where there are no deletions available.

21. Williams syndrome is the result of a deletion of the 7q11.23 region of chromosome 7. Cri du chat syndrome is the result of a deletion of a significant portion of the short arm of chromosome 5 (specifically bands 5p15.2 and 5p15.3). Both Turner syndrome (XO) and Down syndrome (trisomy 21) result from meiotic non-disjunction. The term syndrome is used to describe a set of phenotypes (often complex and varied) that generally occur together.

22. a. Cytologically, deletions lead to shorter chromosomes with missing bands (if banded) and an unpaired loop during meiotic pairing when heterozygous. Genetically, deletions are usually lethal when homozygous, do not revert, and when heterozygous, lower recombinational frequencies and can result in "pseudodominance" (the expression of recessive alleles on one homolog that are deleted on the other). Occasionally, heterozygous deletions express an abnormal (mutant) phenotype.

b. Cytologically, duplications lead to longer chromosomes and, depending on the type, unique pairing structures during meiosis when heterozygous. These may be simple unpaired loops or more complicated twisted loop structures. Genetically, duplications can lead to asymmetric pairing and unequal crossing-over events during meiosis, and duplications of some regions can produce specific mutant phenotypes.

c. Cytologically, inversions can be detected by banding, and when heterozygous, they show the typical twisted "inversion" loop during homologous pairing. Pericentric inversions can result in a change in the p:q ratio (the position of the centromere). Genetically, no viable crossover products are seen from recombination within the inversion when heterozygous, and as a result, flanking genes show a decrease in RF.

d. Cytologically, reciprocal translocations may be detected by banding, or they may drastically change the size of the involved chromosomes as well as the positions of their centromeres. Genetically, they establish new linkage relationships. When heterozygous, they show the typical cross structure during meiotic pairing and cause a diagnostic 50 percent reduction of viable gamete production, leading to semisterility.

23. **a.** paracentric inversion

b. deletion

c. pericentric inversion

d. duplication

24. **a.** The products of crossing-over within the inversion will be inviable when the inversion is heterozygous. This paracentric inversion spans 25 percent of the region between the two loci and therefore will reduce the observed recombination between these genes by a similar percentage (i.e., 9 percent.) The observed RF will be 27 percent.

b. When the inversion is homozygous, the products of crossing-over within the inversion will be viable, so the observed RF will be 36 percent.

25. The following represents the crosses that are described in this problem:

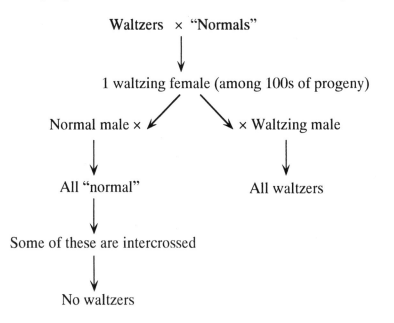

The single waltzing female that arose from a cross between waltzers and normals is expressing a recessive gene. It is possible that this represents a new "waltzer" mutation that was inherited from one of the "normal" mice, but given the cytological evidence (the presence of a shortened chromosome), it is more likely that this exceptional female inherited a deletion of the wild-type allele, which allowed expression of the mutant recessive phenotype.

When this exceptional female was mated to a waltzing male, all the progeny were waltzers; when mated to a normal male, all the progeny were normal. When some of these normal offspring were intercrossed, there were no progeny that were waltzers. If a "new" recessive waltzer allele had been inherited, all these "normal" progeny would have been w^+/w. Any intercross should have therefore produced 25 percent waltzers. On the other hand, if a deletion had occurred, half the progeny would be w^+/w and half would be $w^+/w^{deletion}$. If $w^+/w^{deletion}$ are intercrossed, 25 percent of the progeny would not develop (the homozygous deletion would likely be lethal), and no waltzers would be observed. This is consistent with the data.

26. This problem uses a known set of overlapping deletions to order a set of mutants. This is called deletion mapping and is based on the expression of the recessive mutant phenotype when heterozygous with a deletion of the corresponding allele on the other homolog. For example, mutants *a*, *b*, and *c* are all expressed when heterozygous with Del1. Thus it can be assumed that these genes are deleted in Del1. When these results are compared with the crosses with Del2 and it is discovered that these progeny are b^+, the location of gene *b* is mapped to the region deleted in Del1 that is not deleted in Del2. This logic can be applied in the following way:

Compare deletions 1 and 2: this places allele *b* more to the left than alleles *a* and *c*.
The order is *b* (*a*, *c*), where the parentheses indicate that the order is unknown.

Compare deletions 2 and 3: this places allele *e* more to the right than (*a*, *c*).
The order is *b* (*a*, *c*) *e*.

Compare deletions 3 and 4: allele *a* is more to the left than *c* and *e*, and *d* is more to the right than *e*. The order is *b a c e d*.

Compare deletions 4 and 5: allele *f* is more to the right than *d*. The order is *b a c e d f*.

Allele	Band
b	1
a	2
c	3
e	4
d	5
f	6

27. The data suggest that one or both breakpoints of the inversion are located within an essential gene, causing a recessive lethal mutation.

28. **a.** The Sumatra chromosome contains a pericentric inversion when compared with the Borneo chromosome.

b.

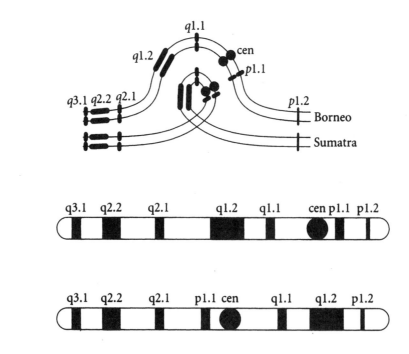

c.

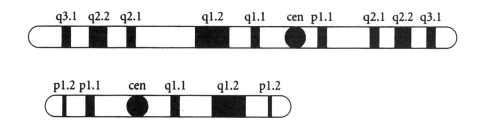

d. Recall that all single crossovers within the inverted region will lead to four meiotic products: two that will be viable, nonrecombinant (parental) types and two that will be extremely unbalanced, (most likely nonviable), recombinant types. In other words, if 30 percent of the meioses have a crossover in this region, 15 percent of the gametes will not lead to viable progeny. That means that 85 percent of the gametes should produce viable progeny.

29. *Unpacking the Problem*

1. A "gene for tassel length" means that there is a gene with at least two alleles (T and t) that controls the length of the tassel. A "gene for rust resistance" means that there is a gene that determines whether the corn plant is resistant to a rust infection or not (R and r).

2. The precise meaning of the allelic symbols for the two genes is irrelevant to solving the problem, because what is being investigated is the distance between the two genes.

3. A locus is the specific position occupied by a gene on a chromosome. It is implied that gene loci are the same on both homologous chromosomes. The gene pair can consist of identical or different alleles.

4. Evidence that the two genes are normally on separate chromosomes would have come from previous experiments showing that the two genes independently assort during meiosis.

5. Routine crosses could consist of F_1 crosses, F_2 crosses, backcrosses, and testcrosses.

6. The genotype T/t ; R/r is a double heterozygote, or dihybrid, or F_1 genotype.

7. The pollen parent is the "male" parent that contributes to the pollen tube nucleus, the endosperm nucleus, and the progeny.

8. Testcrosses are crosses that involve a genotypically unknown and a homozygous recessive organism. They are used to reveal the complete

genotype of the unknown organism and to study recombination during meiosis.

9. The breeder was expecting to observe 1 T/t ; R/r:1 T/t ; r/r:1 t/t ; R/r:1 t/t ; r/r.

10. Instead of a 1:1:1:1 ratio indicating independent assortment, the testcross indicated that the two genes were linked, with a genetic distance of $100\%(3 + 5)/210 = 3.8$ map units.

11. The equality and predominance of the first two classes indicate that the parentals were TR/tr.

12. The equality and lack of predominance of the second two classes indicate that they represent recombinants.

13. The gametes leading to this observation were:
    46.7% TR 1.4% Tr
    49.5% tr 2.4% tR

14. 46.7% TR
    49.5% tr

15. 1.4% Tr
    2.4% tR

16. Tr and tR

17. T and R are linked, as are t and r.

18. Two genes on separate chromosomes can become linked through a translocation.

19. One parent of the hybrid plant contained a translocation that linked the T and R alleles and the t and r alleles.

20. A corn cob is a structure that holds on its surface, the seeds that will become the next generation of corn.

21.

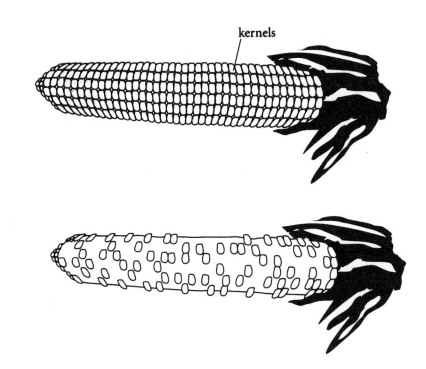

kernels

22.

23. A kernel is one progeny on a corn cob.

24. Absence of half the kernels, or 50 percent aborted progeny (semisterility), could result from the random segregation of one normal with one translocated chromosome (T1 + N2 and T2 + N1) during meioses in a parent that is heterozygous for a reciprocal translocation.

25. Approximately 50 percent of the progeny died. It was the "female" that was heterozygous for the translocation.

Solution to the Problem

a. The progeny are not in the 1:1:1:1 ratio expected for independent assortment; instead, the data indicate close linkage. And, half the progeny did not develop, indicating semisterility.

b. These observations are best explained by a translocation of material between the two chromosomes.

c.   Parents: TR/tr × t/t; r/r

        Progeny: 98 TR/t; r
               104   tr/t; r
                3   Tr/t; r
                5   tR/t; r

d. Assume a translocation heterozygote in coupling. If pairing is as diagrammed below, then you would observe the following:

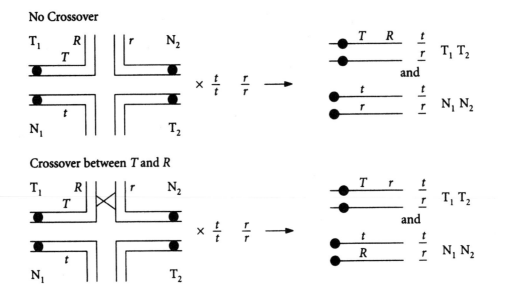

The locations of the genes on the chromosomes is not critical, as long as the cumulative distance between the genes and the breakpoints allows approximately 3.8% recombination.

e. The two recombinant classes result from a recombination event followed by proper segregation of chromosomes, as diagrammed above.

30. The cross was

P X^{e^+}/Y (irradiated) × X^e/X^e

F_1 Most X^e/Y yellow males

Two ? gray males

a. Gray male 1 was crossed with a yellow female, yielding yellow females and gray males, which is reversed sex linkage. If the e^+ allele was translocated to the Y chromosome, the gray male would be X^e/Y^{e^+} or gray. When crossed with yellow females, the results would be

X^e/Y^{e^+} gray males

X^e/X^e yellow females

b. Gray male 2 was crossed with a yellow female, yielding gray and yellow males and females in equal proportions. If the e^+ allele was translocated to an autosome, the progeny would be as below, where "A" indicates autosome:

P $\quad$ $A^{e^+}/A\,;X^e/Y \times A/A\;X^e/X^e$

F$_1$ $\quad$ $A^{e^+}/A\,;X^e/X^e$ gray female

$\quad\quad$ $A^{e^+}/A\,;X^e/Y$ gray male

$\quad\quad$ $A/A\,;X^e/X^e$ yellow female

$\quad\quad$ $A/A\,;X^e/Y$ yellow male

31. The break point can be treated as a gene with two "alleles," one for normal fertility and one for semisterility. The problem thus becomes a two-point cross.

Parentals	764	Semisterile *Pr*
	727	Normal *pr*
Recombinants	145	Semisterile *pr*
	186	Normal *Pr*
	1822	

$100\%(145 + 186)/1822 = 18.17$ m.u.

32.

Klinefelter syndrome	XXY male
Down syndrome	Trisomy 21
Turner syndrome	XO female

33. Create a hybrid by crossing the two plants and then double the chromosomes with a treatment that disrupts mitosis, such as colchicine treatment. Alternatively, diploid somatic cells from the two plants could be fused and then grown into plants through various culture techniques.

34. a.

$\quad\quad\quad\quad b^+/b/b \quad \times \quad b/b$

$\quad\quad\quad\quad\quad \downarrow \quad\quad\quad\quad\quad \downarrow$

Gametes: $\quad \frac{1}{6}\ b^+ \quad\quad\quad b$

$\quad\quad\quad\quad\quad \frac{1}{3}\ b$

$\quad\quad\quad\quad\quad \frac{1}{3}\ b^+/b$

$\quad\quad\quad\quad\quad \frac{1}{6}\ b/b$

Among the progeny of this cross, the phenotypic ratio will be 1 wild-type (b^+) : 1 *b*.

b.

$\quad\quad\quad\quad b^+/b^+/b \quad \times \quad b/b$

$\quad\quad\quad\quad\quad \downarrow \quad\quad\quad\quad\quad \downarrow$

Gametes: $\quad \frac{1}{6}\ b \quad\quad\quad b$

$\quad\quad\quad\quad\quad \frac{1}{3}\ b^+$

$\quad\quad\quad\quad\quad \frac{1}{3}\ b^+/b$

$\quad\quad\quad\quad\quad \frac{1}{6}\ b^+/b^+$

Among the progeny of this cross, the phenotypic ratio will be 5 wild-type (b^+) : 1 b.

c.
$$b^+/b^+/b \quad \times \quad b^+/b$$
$$\downarrow \qquad\qquad \downarrow$$

Gametes:
$1/6$ b $\qquad$ $1/2$ b
$1/3$ b^+ $\qquad$ $1/2$ b^+
$1/3$ b^+/b
$1/6$ b^+/b^+

Among the progeny of this cross, the phenotypic ratio will be 11 wild-type (b^+) : 1 b.

35. **a., b. and c.** One of the parents of the woman with Turner syndrome (XO) must have been a carrier for colorblindness, an X-linked recessive disorder. Because her father has normal vision, she could not have obtained her only X from him. Therefore, nondisjunction occurred in her father. A sperm lacking a sex chromosome fertilized an egg with the X chromosome carrying the colorblindness allele. The nondisjunctive event could have occurred during either meiotic division.

   d. If the colorblind patient had Klinefelter syndrome (XXY), then both Xs must carry the allele for colorblindness. Therefore, nondisjunction had to occur in the mother. Remember that during meiosis I, given no crossover between the gene and the centromere, allelic alternatives separate from each other. During meiosis II, identical alleles on sister chromatids separate. Therefore, assuming there have been no crossovers between the colorblind allele and the centromere, the non-disjunctive event had to occur during meiosis II because both alleles are identical. If the gene is far from the centromere, it would be difficult to determine if non-disjunction happened at MI or MII without molecular studies on haplotypes near the centromere.

36. **a.** If a 6x were crossed with a 4x, the result would be 5x.

   b. Cross A/A with $a/a/a/a$ to obtain $A/a/a$.

   c. The easiest way is to expose the A/a^* plant cells to colchicine for one cell division. This will result in a doubling of chromosomes to yield $A/A/a^*/a^*$.

   d. Cross 6x ($a/a/a/a/a/a$) with 2x (A/A) to obtain $A/a/a/a$.

37. The following answers make the simplifying assumption that there are no crossovers. Without this assumption, it would be hard to tell which stage of meiosis led to the nondisjunction.

Type a: the extra chromosome must be from the mother. Because the chromosomes are identical, nondisjunction had to have occurred at M_{II}.

Type b: the extra chromosome must be from the mother. Because the chromosomes are not identical, nondisjunction had to have occurred at M_I.

Type c: the mother correctly contributed one chromosome, but the father did not contribute any chromosome 4. Therefore, nondisjunction occurred in the male during either meiotic division.

38. **a.** The cross is $P/P/p \times p/p$.

The gametes from the trisomic parent will occur in the following proportions:

$$\begin{array}{ll} 1/6 & p \\ 2/6 & P \\ 1/6 & P/P \\ 2/6 & P/p \end{array}$$

Only gametes that are p can give rise to potato leaves, because potato is recessive. Therefore, the ratio of normal to potato will be 5:1.

b. If the gene is not on chromosome 6, there should be a 1:1 ratio of normal to potato.

39. The generalized cross is $A/A/A \times a/a$, from which $A/A/a$ progeny were selected. These progeny were crossed with a/a individuals, yielding the results given. Assume for a moment that each allele can be distinguished from the other, and let $1 = A$, $2 = A$ and $3 = a$. The gametic combinations possible are

$$\begin{array}{l} \text{1-2 } (A/A) \text{ and 3 } (a) \\ \text{1-3 } (A/a) \text{ and 2 } (A) \\ \text{2-3 } (A/a) \text{ and 1 } (A) \end{array}$$

Because only diploid progeny were examined in the cross with a/a, the progeny ratio should be 2 wild type:1 mutant if the gene is on the trisomic chromosome. With this in mind, the table indicates that y is on chromosome 1, *cot* is on chromosome 7, and h is on chromosome 10. Genes d and c do not map to any of these chromosomes.

40. a.

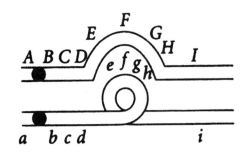

b.

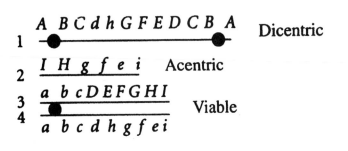

c.

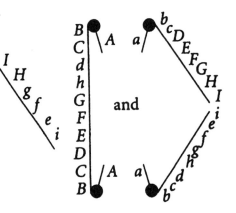

d. The chromosomes numbered 3 and 4 will give rise to viable progeny. The genotypes of those progeny will be *A B C D E F G H I/a b c D E F G H I* and *A B C D E F G H I/a b c d h g f e i*.

41. a. Two crosses show 28 map units between the loci for body color and eye shape in a testcross of the F_1: California × California and Chile × Chile. The third type of cross, California × Chile, leads to only four map units between the two genes when the hybrid is testcrossed. This indicates that the genetic distance has decreased by 24 map units, or $100\%(24/28) = 85.7\%$. A deletion cannot be used to explain this finding, nor can a translocation. Most likely the two lines are inverted with respect to each other for 85.7 percent of the distance between the two genes.

b.

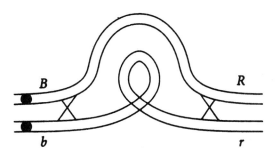

A single crossover in either region would result in 4% crossing-over between *B* and *R*. The products are

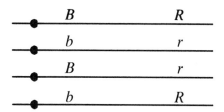

42. **a.** The aberrant plant is semisterile, which suggests an inversion. Because the *d–f* and *y–p* frequencies of recombination in the aberrant plant are normal, the inversion must involve *b* through *x*.

b. To obtain recombinant progeny when an inversion is involved, either a double crossover occurred within the inverted region or single crossovers occurred between *f* and the inversion, which occurred someplace between *f* and *b*.

43. The cross is

P   *c bz wx sh d/c bz wx sh d* × *C Bz Wx Sh D/C Bz Wx Sh D*

F_1 *C Bz Wx Sh D/c bz wx sh d*

Backcross   *C Bz Wx Sh D/c bz wx sh d* × *c bz wx sh d/c bz wx sh d*

a. The total number of progeny is 1000. Classify the progeny as to where a crossover occurred for each type. Then, total the number of crossovers between each pair of genes. Calculate the observed map units.

Region	#COs	M.U. observed	M.U. expected
C–Bz	103	10.3	12
Bz–Wx	13	1.3	8
Wx–Sh	13	1.3	10
A–D	186	18.6	20

Notice that a reduction of map units, or crossing-over, is seen in two intervals. Results like this are suggestive of an inversion. The inversion most likely involves the Bz, Wx, and Sh genes.

Further notice that all those instances in which crossing-over occurred in the proposed inverted region involved a double crossover. This is the expected pattern.

b. A number of possible classes are missing: four single-crossover classes resulting from crossing-over in the inverted region, eight double-crossover classes involving the inverted region and the noninverted region, and triple crossovers and higher. The 10 classes detected were the only classes that were viable. They involved a single crossover outside the inverted region or a double crossover within the inverted region.

c. Class 1: parental; increased due to nonviability of some crossovers

Class 2: parental; increased due to nonviability of some crossovers

Class 3: crossing-over between *C* and *Bz*; approximately expected frequency

Class 4: crossing-over between *C* and *Bz*, approximately expected frequency

Class 5: crossing-over between *Sh* and *D*; approximately expected frequency

Class 6: crossing-over between *Sh* and *D*; approximately expected frequency

Class 7: double crossover between *C* and *Bz* and between *Sh* and *D*; approximately expected frequency

Class 8: double crossover between *C* and *Bz* and between *A* and *D*; approximately expected frequency

Class 9: double crossover between *Bz* and *Wx* and between *Wx* and *Sh*; approximately expected frequency

Class 10: double crossover between *Bz* and *Wx* and between *Wx* and *Sh*; approximately expected frequency

d. Cytological verification could be obtained by looking at chromosomes during meiotic pairing. Genetic verification could be achieved by mapping these genes in the wild-type strain and observing their altered relationships.

44. The original plant was homozygous for a translocation between chromosomes 1 and 5, with breakpoints very close to genes *P* and *S* Because of the close linkage, a ratio suggesting a monohybrid cross, instead of a dihybrid cross, was observed, both with selfing and with a testcross. All gametes are fertile because of homozygosity.

original plant: *P S/p s*
tester: *p s/p s*

F₁ progeny: heterozygous for the translocation:

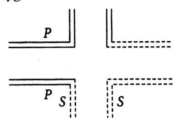

This figure is an example of one configuration that fits the data. One way to test for the presence of a translocation is to look at the chromosomes of heterozygotes during meiosis I.

45. The percent degeneration seen in the progeny of the exceptional rat is roughly 50 percent larger than that seen in the progeny from the normal male. Fifty percent emisterility is an important diagnostic for translocation heterozygotes. This could be verified by cytological observation of the meiotic cells from the exceptional male.

46. **a.** The cross is *Fr/Fr* × *fr/fr/fr*.

Trisomic progeny are then crossed to a diploid wild-type plant.

Fr/fr/fr × *fr/fr*

Because only diploid progeny of this cross are evaluated, the ratio of fast- to slow-ripening plants will be 1:2.

b. If *Fr* is not located on the trisomic chromosome, the crosses are

Fr/Fr × *fr/fr*

and

Fr/fr × *fr/fr*

Therefore the ratio of fast- to slow-ripening plants will be 1:1.

c. The 1:2 ratio of fast- to slow ripening plants indicates that the *Fr* gene is on chromosome 7.

CHALLENGING PROBLEMS

47. a. Single crossovers between a gene and its centromere lead to a tetratype (second-division segregation.) Thus a total of 20 percent of the asci should show second division segregation, and 80 percent will show first-division segregation. The following are representative asci

un3+ ad3+	*un3+ ad3+*	*un3+ ad3+*	*un3+ ad3*	*un3+ ad3*
un3+ ad3+	*un3+ ad3+*	*un3+ ad3+*	*un3+ ad3*	*un3+ ad3*
un3+ ad3+	*un3+ ad3*	*un3+ ad3*	*un3+ ad3+*	*un3+ ad3+*
un3+ ad3+	*un3+ ad3*	*un3+ ad3*	*un3+ ad3+*	*un3+ ad3+*
un3 ad3	*un3 ad3+*	*un3 ad3*	*un3 ad3+*	*un3 ad3*
un3 ad3	*un3 ad3+*	*un3 ad3*	*un3 ad3+*	*un3 ad3*
un3 ad3	*un3 ad3*	*un3 ad3+*	*un3 ad3*	*un3 ad3+*
un3 ad3	*un3 ad3*	*un3 ad3+*	*un3 ad3*	*un3 ad3+*
80%	5%	5%	5%	5%

In all cases, the "upside-down" version would be equally likely.

b. The aborted spores could result from a crossing-over event within an inversion of the wild type, compared with the standard strain. Crossing-over within heterozygous inversions leads to unbalanced chromosomes and nonviable spores. This could be tested by using the wild type from Hawaii in mapping experiments of other markers on chromosome 1 in crosses with the standard strain and looking for altered map distances.

48. Cross 1: independent assortment of the 2 genes (expected for genes on separate chromosomes).

Cross 2: the 2 genes now appear to be linked (the observed RF is 1%); also, half of the progeny are inviable. These data suggest a reciprocal translocation occurred and both genes are very close to the breakpoints.

Cross 3: the viable spores are of 2 types: half contain the normal (nontranslocated chromosomes) and half contain the translocated chromosomes.

49. a.
$$a^+/a^+/a/a \ \times \ a/a/a/a$$
$$\downarrow \qquad\qquad \downarrow$$
Gametes: $\frac{1}{6} \ a^+/a^+$ $\qquad$ a/a

$^2/_3$ a^+/a

$^1/_6$ a/a

Among the progeny of this cross, the phenotypic ratio will be 5 wild-type (a^+) : 1 a.

b.

$$a^+/a/a/a \quad \times \quad a/a/a/a$$
$$\downarrow \qquad\qquad \downarrow$$

Gametes: $^1/_2$ a^+/a \qquad a/a

$^1/_2$ a/a

Among the progeny of this cross, the phenotypic ratio will be 1 wild-type (a^+) : 1 a.

c.

$$a^+/a/a/a \quad \times \quad a^+/a/a/a$$
$$\downarrow \qquad\qquad \downarrow$$

Gametes: $^1/_2$ a^+/a \qquad $^1/_2$ a^+/a

$^1/_2$ a/a \qquad $^1/_2$ a/a

Among the progeny of this cross, the phenotypic ratio will be 3 wild-type (a^+) : 1 a.

d.

$$a^+/a^+/a/a \quad \times \quad a^+/a/a/a$$
$$\downarrow \qquad\qquad \downarrow$$

Gametes: \quad $^1/_6$ a^+/a^+ \qquad $^1/_2$ a^+/a

$^2/_3$ a^+/a \qquad $^1/_2$ a/a

$^1/_6$ a/a

Among the progeny of this cross, the phenotypic ratio will be 11 wild-type (a^+) : 1 a.

50. Consider the following table, in which "L" and "S" stand for 13 large and 13 small chromosomes, respectively:

Hybrid	Chromosomes
G. hirsutum × *G. thurberi*	S, S, L
G. hirsutum × *G. herbaceum*	S, L, L
G. thurberi × *G. herbaceum*	S, L

Each parent in the cross must contribute half its chromosomes to the hybrid offspring. It is known that *G. hirsutum* has twice as many chromosomes as the other two species. Furthermore, because the *G. hirsutum* chromosomes form bivalent pairs in the hybrids, the *G. hirsutum* karyotype must consist of chromosomes donated by the other two species. Therefore, the genome of *G. hirsutum* must consist of one large and one small set of chromosomes. Once this is realized, the rest of the problem essentially solves itself. In the first hybrid, the genome of *G. thurberi* must consist of one set of small chromosomes. In the second hybrid, the genome of *G. herbaceum* must consist of one set of large chromosomes. The third hybrid confirms the conclusions reached from the first two hybrids.

The original parents must have had the following chromosome constitution:

G. hirsutum	26 large, 26 small
G. thurberi	26 small
G. herbaceum	26 large

G. hirsutum is a polyploid derivative of a cross between the two Old World species. This could easily be checked by looking at the chromosomes.

51. a. *B. campestris* was crossed with *B. napus*, and the hybrid had 29 chromosomes consisting of 10 bivalents; and 9 univalents. *B. napus* had to have contributed a total of 19 chromosomes to the hybrid. Therefore, *B. campestris* had to have contributed 10 chromosomes. The *2n* number of *B. campestris* is 20.

When *B. nigra* was crossed with *B. napus*, *B. nigra* had to have contributed 8 chromosomes to the hybrid. The *2n* number of *B. nigra* is 16.

B. oleracea had to have contributed 9 chromosomes to the hybrid formed with *B. juncea*. The *2n* number in *B. oleracea* is 18.

b. First list the haploid and diploid number for each species:

Species	Haploid	Diploid
B. nigra	8	16
B. oleracea	9	18
B. campestris	10	20
B. carinata	17	34
B. juncea	18	36
B. napus	19	38

Now, recall that a bivalent in a hybrid indicates that the chromosomes are essentially identical. Therefore, the more bivalents formed in a hybrid, the closer the two parent species. Three crosses result in no bivalents, suggesting that the parents of each set of hybrids are not closely related:

Cross	Haploid number
B. juncea × *B. oleracea*	18 vs. 9
B. carinata × *B. campestris*	17 vs. 10
B. napus × *B. nigra*	19 vs. 8

Three additional crosses resulted in bivalents, suggesting a closer relationship among the parents:

Cross	Haploid #	Bivalents	Univalents
B. juncea × *B. nigra*	18 vs. 8	8	10
B. napus × *B. campestris*	19 vs. 10	10	9
B. carinata × *B. oleracea*	17 vs. 9	9	8

Note that in each cross the number of bivalents is equal to the haploid number of one species. This suggests that the species with the larger haploid number is a hybrid composed of the second species and some other species. In each case, the haploid number of the unknown species is the number of univalents. Therefore, the following relationships can be deduced:
B. juncea is an amphidiploid formed by the cross of *B. nigra* and *B. campestris*.

B. napus is an amphidiploid formed by the cross of *B. campestris* and *B. oleracea*.

B. carinata is an amphidiploid formed by the cross of *B. nigra* and *B. oleracea*.

These conclusions are in accord with the three crosses that did not yield bivalents:

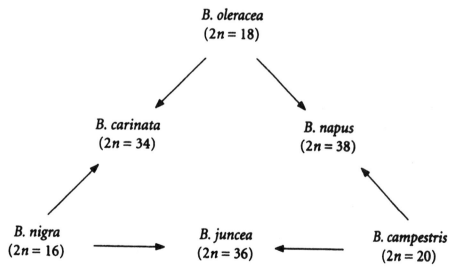

52. a.   Loss of one X in the developing fetus after the two-cell stage

   b.   Nondisjunction leading to Klinefelter syndrome (XXY), followed by a nondisjunctive event in one cell for the Y chromosome after the two-cell stage, resulting in XX and XXYY

c. Nondisjunction of the X at the one-cell stage

d. Fused XX and XY zygotes (from the separate fertilizations either of two eggs or of an egg and a polar body by one X-bearing and one Y-bearing sperm)

e. Nondisjunction of the X at the two-cell stage or later

53. *Unpacking the Problem*

1. *Homozygous* means that an organism has two identical alleles.
   A *mutation* is any deviation from wild type.
   An *allele* is one particular form of a gene.
   Closely linked means two genes are almost always transmitted together through meiosis.
   Recessive refers to a type of allele that is expressed only when it is the sole type of allele for that gene found in an individual.
   Wild type is the most frequent type found in a laboratory population or in a population in the "wild."
   Crossing-over refers to the physical exchange of alleles between homologous chromosomes.
   Nondisjunction is the failure of separation of either homologous chromosomes or sister chromatids in the two meiotic divisions.
   A *testcross* is a cross to a homozygous recessive organism for the trait or traits being studied.
   Phenotype is the appearance of an organism.
   Genotype is the genetic constitution of an organism.

2. No. The genes in question are on an autosome, specifically, number 4.

3. The most common lab species, *Drosophila melanogaster*, has eight chromosomes.

4.   P $b\,e^+/b\,e^+ \times b^+\,e/b^+\,e$ (cross 1)

        F_1 $b\,e^+/b^+\,e \times b\,e/b\,e$ (cross 2)

Progeny $b\,e^+/b\,e$ expected parental
          $b^+\,e/b\,e$ expected parental
          $b^+\,e^+/b\,e$ unexpected recombinant ("very closely linked" so rare)
          $b\,e/b\,e$ unexpected recombinant ("very closely linked" so rare)

          rare wild type × $b\,e/b\,e$ (cross 3)

Progeny	$1/6$	wild type
	$1/6$	bent, eyeless
	$1/3$	bent
	$1/3$	eyeless

5. $b\,e^+$ and $b^+\,e$

6. $b\,e^+/b^+\,e$

7. It is not at all surprising that the F_1 are wild type. This means that both mutations are recessive and complement (are in different genes).

8. $b\,e/b\,e \rightarrow$ gametes: $b\,e$

9. The two common gametes are $b\,e^+$ and $b^+\,e$. The two rare gametes are $b^+\,e^+$ and $b\,e$.

10. Normal

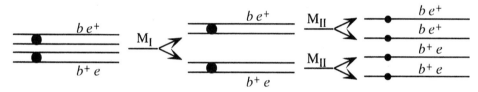

First-Division Nondisjunction: all gametes are aneuploid

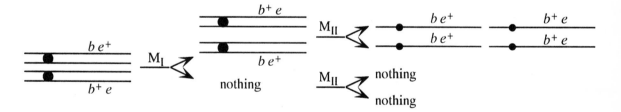

Second-Division Nondisjunction: half the gametes are aneuploid

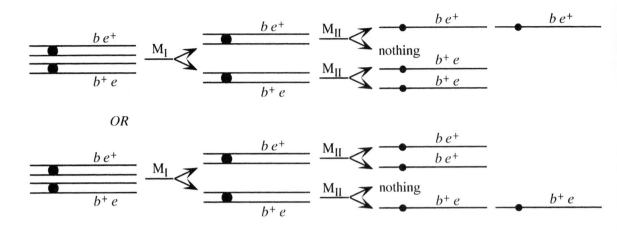

11. This is answered in 10, above.

12. Viable progeny may be able to arise from aneuploidic gametes because chromosome 4 is very small and, percentage-wise, contributes little to the genome. The progeny would be monosomic and trisomic.

13. Listed below are the gametes from 9 and 10 above, the contribution of the male parent, and the phenotype of the progeny.

Female gamete	Male gamete	Phenotype
$b^+ e$	$b\ e$	eyeless
$b\ e^+$	$b\ e$	bent
$b^+ e^+$	$b\ e$	wild type
$b\ e$	$b\ e$	bent, eyeless
$b^+ e/b\ e^+$	$b\ e$	wild type
—	$b\ e$	bent, eyeless
$b\ e^+/b\ e^+$	$b\ e$	bent
$b^+ e/b^+ e$	$b\ e$	eyeless

14. The ratio points to meiosis of a trisomic.

15. Research with artificial chromosomes has indicated that extremely small chromosomes segregate improperly at higher rates than longer chromosomes. It is suspected that the chromatids from homologous chromosomes need to intertwine in order to remain together until the onset of anaphase. Very short chromosomes are thought to have some difficulty in doing this and therefore have a higher rate of nondisjunction. In this instance, which deals with natural chromosomes as opposed to artificial chromosomes, very small chromosomes would be expected to have very little genetic material in them, and therefore their loss or gain may not be of too much importance during development.

16. rare wild type $\times$ $e\,b/e\,b$ (cross 3)

If the rare wild type is from recombination, then the cross becomes

$$b^+\,e^+/b\,e \times b\,e/b\,e$$

Progeny $b^+\,e^+/b\,e$ parental: wild type
                   $b\,e/b\,e$ parental: bent, eyeless
                   $b^+\,e/b\,e$ rare recombinant: eyeless
                   $b\,e^+/b\,e$ rare recombinant: bent

If the rare wild type is from nondisjunction, then the cross becomes

$$b\,e^+/b^+\,e/b\,e \times b\,e/b\,e$$

Progeny $b\,e^+/b^+\,e/b\,e$ wild type
                  $b\,e^+/b\,e/b\,e$ bent
                  $b^+\,e/b\,e/b\,e$ eyeless
                  $b\,e^+/b\,e$ bent
                  $b^+\,e/b\,e$ eyeless
                  $b\,e/b\,e$ bent, eyeless

Solution to the Problem

Cross 1:   P $b\,e^+/b\,e^+$ $\times$ $b^+\,e/b^+\,e$
              F_1 $b^+\,e/b\,e^+$

Cross 2:   P   X/X ; $b^+\,e/b\,e^+$ $\times$ X/Y ; $b\,e/b\,e$
              F_1 expect 1 $b\,e^+/b\,e$: 1 $b^+\,e/b\,e$, X/X and X/Y
                     one rare observed X/X ; $b^+\,e^+$

a. The common progeny are $b^+\,e/b\,e$ and $b\,e^+/b\,e$.

b. The rare female could have come from crossing-over, which would have resulted in a gamete that was $b^+\,e^+$. The rare female could also have come from nondisjunction that gave a gamete that was $b\,e^+/b^+\,e$. Such a gamete might give rise to viable progeny.

c. If the female had been wild-type ($b^+\,e^+/b\,e$) as a result of crossing-over, her progeny would have been as follows:

      Parental: $b^+\,e^+/b\,e$ wild type (common)
                   $b\,e/b\,e$ bent, eyeless (common)

> Recombinant: $b\,e^+/b\,e$ bent (rare)
>
> $b^+\,e/b\,e$ eyeless (rare)

These expected results are very far from what was observed, so the rare female was not the result of recombination.

If the female had been the product of nondisjunction ($b\,e^+/b^+\,e/b\,e$), her progeny when crossed to $b\,e/b\,e$ would be as follows:

$1/6$	$b^+\,e/b\,e$	eyeless
$1/6$	$b\,e^+/b\,e/b\,e$	bent
$1/6$	$b^+\,e/b\,e/b\,e$	eyeless
$1/6$	$b\,e^+/b\,e$	bent
$1/6$	$b\,e/b\,e$	bent, eyeless
$1/6$	$b\,e^+/b^+\,e/b\,e$	wild type

Overall, 2 bent:2 eyeless:1 bent eyeless:1 wild type

These results are in accord with the observed results, indicating that the female was a product of nondisjunction.

54. Recall that ascospores are haploid. The normal genotype associated with the phenotype of each spore is given below:

1	2	3
$b^+\,f^+$	$b\,f^+$	$b^+\,f$
$b^+\,f^+$	$b\,f^+$	$b^+\,f$
$b^+\,f^+$	abort	$b^+\,f^+$
$b^+\,f^+$	abort	$b^+\,f^+$
abort	$b^+\,f$	$b\,f$
abort	$b^+\,f$	$b\,f$
abort	$b^+\,f$	$b\,f^+$
abort	$b^+\,f$	$b\,f^+$

a. For the first ascus, the most reasonable explanation is that nondisjunction occurred at the first meiotic division. Second-division nondisjunction or chromosome loss are two explanations of the second ascus. Crossing-over best explains the third ascus.

b.

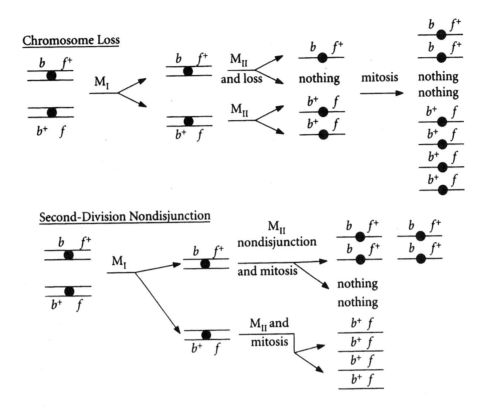

55. a. Each mutant is crossed with wild type, or

$$m \times m^+$$

The resulting tetrads (octads) show 1:1 segregation, indicating that each mutant is the result of a mutation in a single gene.

b. The results from crossing the two mutant strains indicate either both strains are mutant for the same gene,

$$m_1 \times m_2$$

or, that they are mutant in different but closely linked genes

$$m_1 \, m_2{}^+ \times m_1{}^+ m_2$$

c. and d. Because phenotypically black offspring can result from nondisjunction (notice that in Case C and Case D, black appears in conjunction with aborted spores), it is likely that mutant 1 and mutant 2 are mutant in different but closely linked genes. The cross is therefore

$$m_1\, m_2{}^+ \times m_1{}^+ m_2$$

Case A is an NPD tetrad and would be the result of a four-strand double cross-over.

$m_1{}^+\, m_2{}^+$	black
$m_1{}^+\, m_2{}^+$	black
$m_1\, m_2$	fawn
$m_1\, m_2$	fawn

Case B is a tetratype and would be the result of a single cross-over between one of the genes and the centromere.

$m_1{}^+\, m_2{}^+$	black
$m_1{}^+\, m_2$	fawn
$m_1\, m_2{}^+$	fawn
$m_1\, m_2$	fawn

Case C is a the result of nondisjunction during meiosis I.

$m_1{}^+\, m_2\,;\, m_1\, m_2{}^+$	black
$m_1{}^+\, m_2\,;\, m_1\, m_2{}^+$	black
no chromosome	abort
no chromosome	abort

Case D is a the result of recombination between one of the genes and the centromere followed by nondisjunction during meiosis II. For example:

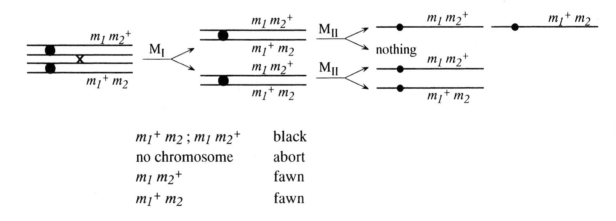

$m_1{}^+\, m_2\,;\, m_1\, m_2{}^+$	black
no chromosome	abort
$m_1\, m_2{}^+$	fawn
$m_1{}^+\, m_2$	fawn

17

Population Genetics

BASIC PROBLEMS

1. The frequency of an allele in a population can be altered by natural selection, mutation, migration, nonrandom mating, and genetic drift (sampling errors).

2. There are a total of $(2)(384) + (2)(210) + (2)(260) = 1708$ alleles in the population. Of those, $(2)(384) + 210 = 978$ are A_1 and $210 + (2)(260) = 730$ are A_2. The frequency of A_1 is $978/1708 = 0.57$, and the frequency of A_2 is $730/1708 = 0.43$.

3. The given data are $q^2 = 0.04$ and $p^2 + 2pq = 0.96$. Assuming Hardy–Weinberg equilibrium, if $q^2 = 0.04$, $q = 0.2$, and $p = 0.8$. The frequency of B/B is $p^2 = 0.64$, and the frequency of B/b is $2pq = 0.32$.

4. No. It proves that the shiny wing phenotype is more common among the beetles observed.

   The frequency of a phenotype in a population is a function of the frequency of alleles that lead to that phenotype in the population. To determine dominance and recessiveness, do standard Mendelian crosses.

5. **a.** The needed equations are

$$p' = p\frac{pW_{AA} + qW_{Aa}}{\overline{W}}$$

and

$$\overline{W} = p^2W_{AA} + 2pqW_{Aa} + q^2W_{aa}$$

$p' = 0.5\,[(0.5)(0.9) + 0.5(1.0)]/[(0.25)(0.9) + (0.5)(1.0) + (0.25)(0.7)] = 0.53$

   b. The needed equation is

$$\hat{p} = \frac{W_{a/a} - W_{A/a}}{(W_{a/a} - W_{A/a}) + (W_{A/A} - W_{A/a})}$$

$$= \frac{0.7 - 1.0}{(0.7 - 1.0) + (0.9 - 1.0)}$$

$$= 0.75$$

6. The needed equation is

$$q^2 = {}^{\mu}/_s$$

or

$$s = \mu/q^2 = 10^{-5}/10^{-3} = 0.01$$

7. Albinos appear to have had greater opportunity to mate. They may have been considered lucky and encouraged to breed at very high levels in comparison with nonalbinos. They may also have been encouraged to mate with each other. Alternatively, in the tribes with a very low frequency, albinos may have been considered very unlucky and destroyed at birth or prevented from marriage.

CHALLENGING PROBLEMS

8. This problem assumes that there is no backward mutation. The problem also assumes that D and d are equally fit. Use the following equation:

$$p_n = p_o e^{-n\mu}$$

That is, $p_{50,000} = (0.8)e^{-(5 \times 10^4)(4 \times 10^{-6})} = (0.8)e^{-0.2} = 0.65$

9. a. If the variants represent different alleles of gene X, a cross between any two variants should result in a 1:1 progeny ratio (because the organism is haploid.) All the variants should map to the same locus. Amino acid sequencing of the variants should reveal differences of one to just a few amino acids.

   b. There could be another gene (gene Y), with five variants, that modifies the gene X product post-transcriptionally. If so, the easiest way to distinguish between the two explanations would be to find another mutation in X and do a dihybrid cross. For example, if there is independent assortment,

   P $X^1 ; Y^1 \times X^2 ; Y^2$
   F_1 $1 X^1 ; Y^1 : 1 X^1 ; Y^2 : 1 X^2 ; Y^1 : 1 X^2 ; Y^2$

If the new mutation in X led to no enzyme activity, the ratio would be

2 no activity : 1 variant one activity : 1 variant two activity
The same mutant in a one-gene situation would yield 1 active : 1 inactive.

10. **a.** If the population is in equilibrium, $p^2 + 2pq + q^2 = 1$. Calculate the actual frequencies of p and q in the population and compare their genotypic distribution to the predicted values. For this population:

$$p = [406 + \frac{1}{2}(744)]/1482 = 0.52$$
$$q = [332 + \frac{1}{2}(744)]/1482 = 0.48$$

The genotypes should be distributed as follows, if the population is in equilibrium:

$L^M/L^M = p^2(1482) = 401$ Actual: 406
$L^M/L^N = 2pq(1482) = 740$ Actual: 744
$L^N/L^N = q^2(1482) = 341$ Actual: 332

This compares well with the actual data, so the population is in equilibrium.

b. If mating is random with respect to blood type, then the following frequency of matings should occur:

$L^M/L^M \times L^M/L^M = (p^2)(p^2)(741) = 54$ Actual: 58
$L^M/L^M \times L^M/L^N$ or $L^M/L^N \times L^M/L^M = (2)(p^2)(2pq)(741) = 200$ Actual: 202
$L^M/L^N \times L^M/L^N = (2pq)(2pq)(741) = 185$ Actual: 190
$L^M/L^M \times L^N/L^N$ or $L^N/L^N \times L^M/L^M = (2)(p^2)(q^2)(741) = 92$ Actual: 88
$L^M/L^N \times L^N/L^N$ or $L^N/L^N \times L^M/L^N = 2(2pq)(q^2)(741) = 170$ Actual: 162
$L^N/L^N \times L^N/L^N = (q^2)(q^2)(741) = 39$ Actual: 41

Again, this compares nicely with the actual data, so the mating is random with respect to blood type.

11. **a. and b.** For each, p and q must be calculated and then compared with the predicted genotypic frequencies of $p^2 + 2pq + q^2 = 1$.

Population	p	q	Equilibrium?
1	1.0	0.0	Yes
2	0.5	0.5	No
3	0.0	1.0	Yes
4	0.625	0.375	No
5	0.375	0.625	No
6	0.5	0.5	Yes
7	0.5	0.5	No
8	0.2	0.8	Yes

9	0.8	0.2	Yes
10	0.993	0.007	Yes

c. The formulas to use are $q^2 = \mu/s$ and $s = 1-W$.

$$4.9 \times 10^{-5} = 5 \times 10^{-6}/s \; ; \; s = 0.102, \text{ so } W = 0.898$$

d. For simplicity, assume that the differences in survivorship occur prior to reproduction. Thus, each genotype's fitness can be used to determine the relative percentage each contributes to the next generation.

Genotype	Frequency	Fitness	Contribution	A	a
A/A	0.25	1.0	0.25	0.25	0.0
A/a	0.50	0.8	0.40	0.20	0.20
a/a	0.25	0.6	0.15	0.0	0.15
				0.45	0.35

$$p' = 0.45/(0.45 + 0.35) = 0.56$$
$$q' = 0.35/(0.45 + 0.35) = 0.44$$

Alternatively, the formulas to use are

$$p' = p\frac{pW_{AA} + qW_{Aa}}{\overline{W}}$$

and

$$\overline{W} = p^2 W_{AA} + 2pqW_{Aa} + q^2 W_{aa}$$

$$p' = (0.5)[(0.5)(1.0) + (0.5)(0.8)]/[(0.25)(1.0) + (0.5)(0.8) + (0.25)(0.6)]$$
$$= (0.5)(0.9)/(0.8) = 0.56$$

12. a. Assuming the population is in Hardy-Weinberg equilibrium and that the allelic frequency is the same in both sexes, we can directly calculate the frequency of the colorblind allele as $q = 0.1$. (Because this trait is sex-linked, q is equal to the frequency of affected males.) Colorblind females must be homozygous for this X-linked recessive trait, so their frequency in the population is equal to $q^2 = 0.01$.

b. There would be 10 colorblind men for every colorblind woman (q/q^2).

c. For this condition to be true, the mothers must be heterozygous for the trait and the fathers must be colorblind ($X^C/X^c \times X^c/Y$). The frequency of heterozygous women in the population will be $2pq$, and the frequency of colorblind men will be q. Therefore, the frequency of such random marriages will be $(2pq)(q) = 0.018$.

d. All children will be phenotypically normal only if the mother is homozygous for the noncolorblind allele ($p^2 = 0.81$). The father's genotype does not matter and therefore can be ignored.

e. There are several ways of approaching this problem. One way to visualize the data, however, is to construct the following

Mother \ Father	0.4 X^C	0.6 X^c	Y
0.8 X^C	0.32 X^C/X^C	0.48 X^C/X^c	0.8 X^C/Y
0.2 X^c	0.08 X^C/X^c	0.12 X^c/X^c	0.2 X^c/Y

As can be seen, the frequency of colorblind females will be 0.12 and colorblind males 0.2.

f. From analysis of the results in (e), the frequency of the colorblind allele will be 0.2 in males (the same as in the females of the previous generation) and $^1/_2(0.08 + 0.48) + 0.12 = 0.4$ in females.

13. Assume that proper function results from the right gene products in the proper ratio to all other gene products. A mutation will change the gene product, eliminate the gene product, or change the ratio of it to all other gene products. All three outcomes upset a previously balanced system. While a new and "better" balance may be achieved, this is less likely than being deleterious.

14. Wild-type alleles are usually dominant because most mutations result in lowered or eliminated function. To be dominant, the heterozygote has approximately the same phenotype as the dominant homozygote. This will typically be true when the wild-type allele produces a product and the mutant allele does not.

Chromosomal rearrangements are often dominant mutations because they can cause gross changes in gene regulation or even cause fusions of several gene products. Novel activities, overproduction of gene products, etc., are typical of dominant mutations.

15. Prior to migration, $q^A = 0.1$ and $q^B = 0.3$ in the two populations. Because the two populations are equal in number, immediately after migration, $q^{A + B} = ^1/_2(q^A + q^B) = ^1/_2(0.1 + 0.3) = 0.2$. At the new equilibrium, the frequency of affected males is $q = 0.2$, and the frequency of affected females is $q^2 = (0.2)^2 = 0.04$. (Colorblindness is an X-linked trait.)

16. For a population in equilibrium, the probability of individuals being homozygous for a recessive allele is q^2. Thus for small values of q, few individuals in a randomly mating population will express the trait. However, if two individuals share a close common ancestor, there is an increased chance of homozygosity by descent, because only one "progenitor" need be heterozygous.

For the following, it is assumed that the allele in question is rare. Thus the chance of both "progenitors" being heterozygous will be ignored.

a. For a parent-sib mating, the pedigree can be represented as follows:

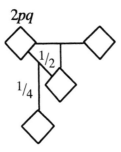

In this example, it is only the chance of the incestuous parent's being heterozygous that matters. Thus, the chance of the descendant's being homozygous is

$$2pq(^1/_2)(^1/_4) = {}^{pq}/_4$$

If q is very small, then p is nearly 1.0 and the chance of an affected child can be represented as approximately $^q/_4$. (Again, this should be compared to the expected random-mating frequency of q^2.)

b. For a mating of first cousins, the pedigree can be represented as follows:

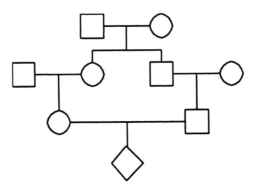

The probability of inheriting the recessive allele if *either* grandparent is heterozygous can be represented as follows:

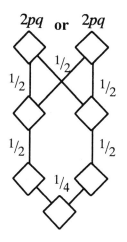

Thus the chance of this child's being affected is

$$2pq\ (^1/_2)(^1/_2)(^1/_2)(^1/_2)(^1/_4) + 2pq\ (^1/_2)(^1/_2)(^1/_2)(^1/_2)(^1/_4) = {pq}/{16}$$

Again, if q is rare, p is nearly 1.0, so the chance of homozygosity by descent is approximately $q/16$.

c.   An aunt-nephew (or uncle-niece) mating can be represented as:

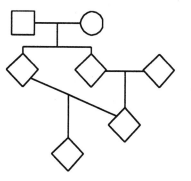

Following the possible inheritance of the recessive allele from either grandparent,

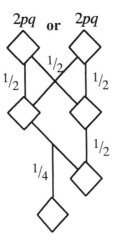

The chance of this child's being homozygous is

$$2pq(^1/_2)(^1/_2)(^1/_2)(^1/_4) + 2pq(^1/_2)(^1/_2)(^1/_2)(^1/_4) = {}^{pq}/_8,$$

or for rare alleles approximately $^q/_8$.

17. The allele frequencies are
$$f(A) = 0.2 + ^1/_2(0.60) = 50\%$$
$$f(a) = ^1/_2(0.60) + 0.2 = 50\%$$

Positive assortative mating: the alleles will randomly unite within the same phenotype. For *A/–*, the mating population is 0.2 *A/A* + 0.6 *A/a*. The allelic frequencies within this subpopulation are

$$f(A) = [0.2 + ^1/_2(0.6)]/0.8 = 0.625$$
$$f(a) = ^1/_2(0.6)/0.8 = 0.375$$

The phenotypic frequencies that result are

A/–: $p^2 + 2pq = (0.625)^2 + 2(0.625)(0.375) = 0.39 + 0.47 = 0.86$
a/a: $q^2 = (0.375)^2 = 0.14$

However, assuming that all contribute equally to the next generation and this subpopulation represents 0.8 of the total population, these figures must be adjusted to reflect this weighting:

A/–: $(0.86)(0.8) = 0.69$
a/a: $(0.14)(0.8) = 0.11$

The *a/a* contribution from the other subpopulation will remain unchanged because there is only one genotype, *a/a*. Their weighted contribution to the total phenotypic frequency is 0.20. Therefore, after one generation, the phenotypic frequencies will be $A/- = 0.69$ and $a/a = 0.20 + 0.11 = 0.31$, and the genotypic frequencies will be $f(A) = 0.5$ and $f(a) = 0.5$. Over time, these allelic frequencies will stay the same, but the frequency of heterozygotes will continue to decrease until there are two separate populations, A/A and a/a, which will not interbreed. *Negative assortative mating*: mating is between unlike phenotypes. The two types of progeny will be A/a and a/a. A/A will not exist. A/a will result from all $A/A \times a/a$ matings and half the $A/a \times a/a$ matings. These matings will occur with the following relative frequencies

$$A/A \times a/a = (0.2)(0.2) = 0.04$$
$$A/a \times a/a = (0.6)(0.2) = 0.12$$

Because these are the only matings that will occur, they must be put on a 100 percent basis by dividing by the total frequency of matings that occur:

$A/A \times a/a$: $0.04/0.16 = 0.25$, all of which will be A/a
$A/a \times a/a$: $0.12/0.16 = 0.75$, half A/a and half a/a

The phenotypic frequencies in this generation will be

A/a: $0.25 + 0.75/2 = 0.625$
a/a: $0.75/2 = 0.375$

In the next generation, because all matings are now between heterozygotes and homozygous recessives, the final allelic frequencies of $f(A) = 0.25$ and $f(a) = 0.75$ will be obtained and the population will be 50 percent A/a and 50 percent a/a.

18. Many genes affect bristle number in *Drosophila*. The artificial selection resulted in lines with mostly high-bristle-number alleles. Some mutations may have occurred during the 20 generations of selective breeding, but most of the response was due to alleles present in the original population. Assortment and recombination generated lines with more high-bristle-number alleles.

Fixation of some alleles causing high bristle number would prevent complete reversal. Some high-bristle-number alleles would have no negative effects on fitness, so there would be no force pushing bristle number back down because of those loci.

The low fertility in the high-bristle-number line could have been due to pleiotropy or linkage. Some alleles that caused high bristle number may also have caused low fertility (pleiotropy). Chromosomes with high-bristle-number alleles may also carry alleles at different loci that caused low fertility (linkage). After artificial selection was relaxed, the low-fertility alleles would have been selected against by natural selection. A few generations of relaxed selection would have allowed low-fertility-linked alleles to recombine away, producing high-bristle-number chromosomes that did not contain low-fertility alleles. When selection was reapplied, the low-fertility alleles had been reduced in frequency or separated from the high-bristle loci, so this time there was much less of a fertility problem.

19. Affected individuals = $B/b = 2pq = 4 \times 10^{-6}$. Because q is almost equal to 1.0, $2p = 4 \times 10^{-6}$. Therefore, $p = 2 \times 10^{-6}$.

$$\mu = hsp = (1.0)(0.7)(2 \times 10^{-6}) = 1.4 \times 10^{-6}$$

where h = degree of dominance of the deleterious allele

20. The probability of not getting a recessive lethal genotype for one gene is $1 - 1/8 = 7/8$. If there are n lethal genes, the probability of not being homozygous for any of them is $(7/8)^n = 13/31$. Solving for n, an average of 6.5 recessive lethals are predicted. If the actual percentage of "normal" children is less owing to missed in utero fatalities, the average number of recessive lethals would be higher.

21. a. The formula needed is
$$\hat{q} = \sqrt{\mu/s}$$
$$= 4.47 \times 10^{-3}$$

so,   Genetic cost = $sq^2 = 0.5(4.47 \times 10^{-3})^2 = 10^{-5}$

b. Using the same formulas as part a,
$$\hat{q} = 6.32 \times 10^{-3}$$
Genetic cost = $sq^2 = 0.5(6.32 \times 10^{-3})^2 = 2 \times 10^{-5}$

c.
$$\hat{q} = 5.77 \times 10^{-3}$$
Genetic cost = $sq^2 = 0.3(5.77 \times 10^{-3})^2 = 10^{-5}$

18

Quantitative Genetics

BASIC PROBLEMS

1. There are many traits that vary more or less continuously over a wide range. For example, height, weight, shape, color, reproductive rate, metabolic activity, etc., vary quantitatively rather than qualitatively. Continuous variation can often be represented by a bell-shaped curve, where the "average" phenotype is more common than the extremes. Discontinuous variation describes the easily classifiable, discrete phenotypes of simple Mendelian genetics: seed shape, auxotrophic mutants, sickle-cell anemia, etc. These traits show a simple relationship between genotype and phenotype.

2. The mean (or average) is calculated by dividing the sum of all measurements by the total number of measurements, or in this case, the total number of bristles divided by the number of individuals.

$$\text{mean} = \bar{x} = \frac{[1 + 4(2) + 7(3) + 31(4) + 56(5) + 17(6) + 4(7)]}{(1 + 4 + 7 + 31 + 56 + 17 + 4)}$$

$$= {}^{564}/_{120} = 4.7 \text{ average number of bristle/individual}$$

The variance is useful for studying the distribution of measurements around the mean and is defined in this example as

$$\text{variance} = s^2 = \text{average of the (actual bristle count} - \text{mean})^2$$

$$\begin{aligned} s^2 &= {}^1/_N \, \Sigma \, (x_i - \bar{x})^2 \\ &= {}^1/_{120} \Sigma [(1 - 4.7)^2 + (2 - 4.7)^2 + (3 - 4.7)^2 + (4 - 4.7)^2 + (5 - 4.7)^2 + \\ &\quad (6 - 4.7)^2 + (7 - 4.7)^2] \\ &= 0.26 \end{aligned}$$

The standard deviation, another measurement of the distribution, is simply calculated as the square root of the variance.

$$\text{standard deviation} = s = \sqrt{0.26} = 0.51$$

3.   a. H^2 has meaning only with respect to the population that was studied in the environment in which it was studied. Even if a trait shows high heritability, it does not imply the trait is unaffected by its environment. The only acceptable analysis is to study directly the norms of reaction of the various genotypes in the population over the range of projected environments. Because it is so difficult to fully replicate a human genotype so that it might be tested in different environments, there is no known norm of reaction for any human quantitative trait.

      b.   Neither H^2 nor h^2 is a reliable measure that can be used to generalize from a particular sample to a "universe" of the human population. They certainly should not be used in social decision making (as implied by the terms eugenics and dysgenics).

      c.   Again, H^2 and h^2 are not reliable measures, and they should not be used in any decision making with regard to social policies.

4.   The following are unknown: (1) norms of reaction for the genotypes affecting IQ, (2) the environmental distribution in which the individuals developed, and (3) the genotypic distributions in the populations. Even if the above were known, because heritability is specific to a specific population and its environment, the difference between two different populations cannot be given a value of heritability.

CHALLENGING PROBLEMS

5.   a.   Broad heritability measures that portion of the total variance that is due to genetic variance. The equation to use is

H^2 = the genetic variance/phenotypic variance

where genetic variance = phenotypic variance − environmental variance

$$H^2 = \frac{s_p^2 - s_e^2}{s_p^2}$$

Narrow heritability measures that portion of the total variance that is due to the additive genetic variation. The equation to use is

$$h^2 = \frac{\text{additive genetic variance}}{\text{additive genetic variance} + \text{dominance variance} + \text{environmental variance}}$$

$$h^2 = \frac{s_a^2}{s_a^2 + s_d^2 + s_e^2}$$

Shank length:

$H^2 = (310.2 - 248.1)/(310.2) = 0.200$

$h^2 = 46.5/(46.5 + 15.6 + 248.1) = 0.150$

Neck length:

$H^2 = (730.4 - 292.2/(730.4) = 0.600$

$h^2 = 73.0/(73.0 + 365.2 + 292.2) = 0.010$

Fat content:

$H^2 = (106.0 - 53.0)/(106.0) = 0.500$

$h^2 = 42.4/(42.4 + 10.6 + 53.0) = 0.400$

b. The larger the value of h^2, the greater the difference between selected parents and the population as a whole, and the more that characteristic will respond to selection. Therefore, fat content would respond best to selection.

c. The formula needed is

selection response = $h^2 \times$ selection differential

Therefore, selection response = $(0.400)(10.5\% - 6.5\%) = 1.6\%$ decrease in fat content, or 8.9% fat content.

6. a. The probability of any gene being homozygous is $1/2$ (e.g., for A: A/A or a/a), and the probability of being heterozygous (or not homozygous) is also $1/2$. Thus, the probability for any one gene being homozygous while the other two are heterozygous is $(1/2)^3$. Because there are three ways for this to happen (homozygosity at A or at B or at C), the total probability is

p(homozygous at 1 locus) = $3(1/2)^3 = 3/8$

The same logic can be applied to any two genes being homozygous

p(homozygous at 2 loci) = $3(1/2)^3 = 3/8$

There are two ways for all three genes to be homozygous, so

$$p(\text{homozygous at 3 loci}) = (1/2)^3 = 1/8$$

b. $p(0 \text{ capital letters}) = p(\text{all homozygous recessive}) = (1/4)^3 = 1/64$

$p(1 \text{ capital letter}) = p(1 \text{ heterozygote and 2 homozygous recessive})$
$= 3(1/2)(1/4)(1/4) = 3/32$

$p(2 \text{ capital letters}) = p(1 \text{ homozygous dominant and 2 homozygous recessive})$

or

$p(2 \text{ heterozygotes and 1 homozygous recessive})$

$$= 3(1/4)^3 + 3(1/4)(1/2)^2 = 15/64$$

$p(3 \text{ capital letters}) = p(\text{all heterozygous})$

or

$p(1 \text{ homozygous dominant, 1 heterozygous, and 1 homozygous recessive})$

$$= (1/2)^3 + 6(1/4)(1/2)(1/4) = 10/32$$

$p(4 \text{ capital letters}) = p(2 \text{ homozygous dominant and 1 homozygous recessive})$

or

$p(1 \text{ homozygous dominant and 2 heterozygous})$

$$= 3(1/4)^3 + 3(1/4)(1/2)^2 = 15/64$$

$p(5 \text{ capital letters}) = p(2 \text{ homozygous dominant and 1 heterozygote})$

$$= 3(1/4)^2(1/2) = 3/32$$

$p(6 \text{ capital letters}) = p(\text{all homozygous dominant}) = (1/4)^3 = 1/64$

7. For three genes there are a total of 27 genotypes that will occur in predictable proportions. For example, there are three genotypes that have two genes that are heterozygous and one gene that is homozygous recessive (A/a ; B/b ; c/c, A/a ; b/b ; C/c, a/a ; B/b ; C/c). The frequency of this combination is $3(1/2)(1/2)(1/4) = 3/16$,

and the phenotypic score is $3 + 3 + 1 = 7$. For all the genotypes possible, the total distribution of phenotypic scores is as follows:

Score	Proportion
3	$1/64$
5	$3/32$
6	$3/64$
7	$3/16$
8	$3/16$
9	$11/64$
10	$3/16$
11	$3/32$
12	$1/64$

And the plot of these data will be

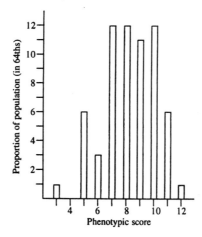

8. The population described would be distributed as follows:

3 bristles	$19/64$
2 bristles	$44/64$
1 bristle	$1/64$

The 3-bristle class would contain seven different genotypes, the 2-bristle class would contain 19 different genotypes, and the 1-bristle class would contain only one genotype. It would be very difficult to determine the underlying genetic situation by doing controlled crosses and determining progeny frequencies.

9. a. Solving the formula for values of x over the stated range for each genotype gives the following data:

x	1	2	3
0.03	0.90		
0.04	0.91		
0.05	0.93		
0.06	0.93		
0.07	0.94		
0.08	0.95		
0.09	0.96		
0.10	0.97		0.90
0.11	0.97		0.92
0.12	0.98	0.90	0.93
0.13	0.98	0.92	0.94
0.14	0.99	0.94	0.95
0.15	0.99	0.95	0.96
0.16	0.99	0.96	0.97
0.17	1.00	0.98	0.98
0.18	1.00	0.98	0.98
0.19	1.00	0.99	0.99
0.20	1.00	1.00	0.99
0.21	1.00	1.00	1.00
0.22	1.00	1.00	1.00
0.23	1.00	1.00	1.00
0.24	0.99	1.00	1.00
0.25	0.99		1.00
0.26	0.99		1.00
0.27	0.98		1.00
0.28	0.98		0.99
0.29	0.97		0.99
0.30	0.97		0.98
0.31	0.96		0.98
0.32	0.95		0.97
0.33	0.94		0.96
0.34	0.93		0.95
0.35	0.93		0.94
0.36	0.91		0.93
0.37	0.90		0.92
0.38			0.90

Plotting these data give the following curves

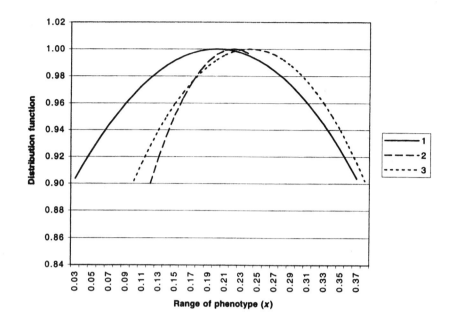

b. Because the three genotypes are equally frequent, the average distribution across the entire range of phenotypes will be

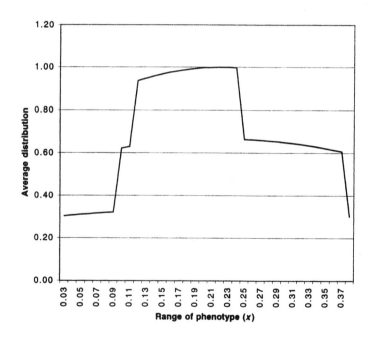

There are regions of the overall phenotypic distribution where the variation within a given genotype does not overlap, and this gives sharp steps to the distribution. On the other hand, any individual whose phenotype lies between values of 0.12 to 0.24 could have any of the three genotypes.

10. **a.**

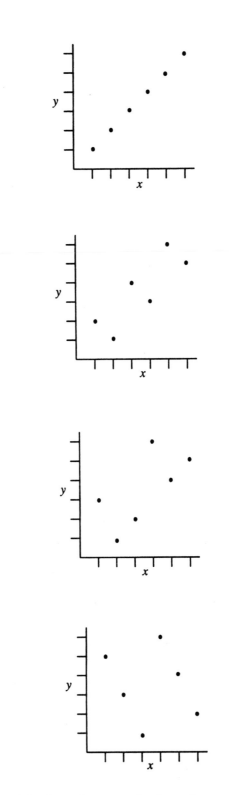

b.

c.

d.

Use the following formulas to calculate the correlation coefficient (r_{xy}) between x and y:

$$r_{xy} = \frac{\text{cov } xy}{s_x s_y}$$

and

$$\text{cov } xy = \frac{1}{N} \Sigma \; x_i \, y_i - \overline{xy}$$

a. $\text{cov } xy = \frac{1}{6} [(1)(1) + (2)(2) + (3)(3) + (4)(4) + (5)(5) + (6)(6)]$
$- (^{21}/_6)(^{21}/_6) = 2.92$

Standard deviation $x = s_x = \sqrt{\frac{1}{N} \Sigma \; (x_i - \overline{x})^2} = 1.71$

Standard deviation $y = s_y = \sqrt{\frac{1}{N} \Sigma \; (y_i - \overline{y})^2} = 1.71$

Therefore, $r_{xy} = 2.92/(1.71)(1.71) = 1.0$. The other correlation coefficients are calculated in a like manner.

b. 0.83

c. 0.66

d. −0.20

11. First, define alcoholism in behavioral terms. Next, realize that all observations must be limited to the behavior you used in the definition and all conclusions from your observations are applicable only to that behavior. In order to do your data gathering, you must work with a population in which familiality is distinguished from heritability. In practical terms, this means using individuals who are genetically close but who are found in all environments possible.

12. Before beginning, it is necessary to understand the data. The first entry, h/h h/h, refers to the II and III chromosomes, respectively. Thus, there are four h (high bristle number) sets of alleles in two or more genes on separate chromosomes. The next entry is h/l h/h. Chromosome II is now heterozygous and chromosome III is still homozygous, etc.

The effect of substituting one l chromosome II for an h chromosome II, and therefore going from homozygous h/h to heterozygous h/l, can be seen in the differences along the rows in the first two columns. The average change is (2.9 + 3.1 + 2.7)/3 = 2.9. When chromosome II goes from heterozygous h/l to homozygous l/l, the average change is (3.2 + 5.2 + 6.8)/3 = 5.1.

The effect of substituting one l chromosome III for an h chromosome III can be seen in the differences between rows: 25.1 − 23.0 = 2.1; 22.2 − 19.9 = 2.3; and 19.0 − 14.7 = 4.3 (average change 2.9). And going from h/l to l/l for chromosome III gives an average change of (11.2 + 10.8 + 12.4)/3 = 11.5 bristles.

Here is a summary of these results:

	Chromosome II	Chromosome III
h/h to h/l	2.9	2.9
h/l to l/l	5.1	11.5

Each set of alleles for both chromosomes is expressed in the phenotype, but that expression varies with the chromosome. Chromosome III appears to have a stronger effect on the phenotype than does chromosome II. (Compare h/h h/h with both l/l h/h and h/h l/l. The difference in the first case is 6.1, and in the second case, 13.3.) Finally, there is partial dominance of h over l for both chromosomes. The change from h/h to h/l is less than the change from h/l to l/l.

13. a.

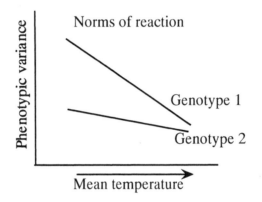

b. Broad heritability is defined as

$$H^2 = \frac{s_g^2}{s_p^2}$$

Assuming that the genetic variance stays the same, phenotypic variance must decrease if H^2 increases. Therefore, the same plot as (a) will satisfy the conditions.

c. To satisfy the conditions that genetic variance increases as H^2 decreases, phenotypic variance must also increase. Therefore the plot will be

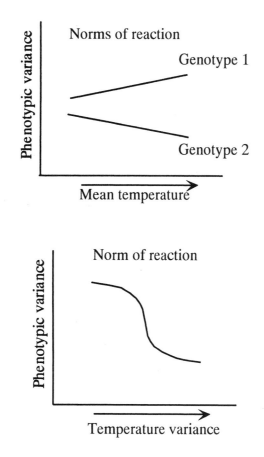

d.

14. a. The regression line shows the relationship between the two variables. It attempts to predict one (the son's height) from the other (the father's height). If the relationship is perfectly correlated, the slope of the regression line should approximate 1. If you assume that individuals at the extreme of any spectrum are homozygous for the genes responsible for these phenotypes, then their offspring are more likely to be heterozygous than are the original individuals. That is, they will be less extreme. Also, there is no attempt to include the maternal contribution to this phenotype.

b. For Galton's data, regression is an estimate of heritability (h^2), *assuming* that there were few environmental differences between all fathers and all sons both, individually and as a group. However, there is no evidence given to determine if the traits are familial but not heritable. This data would indicate genetic variation only if the relatives do not share common environments more than nonrelatives do.

19

Evolutionary Genetics

BASIC PROBLEMS

1. A transformational scheme of evolution predicts that all members of a species will change over time. By sharing similar environments and life experiences, each member is altered in its lifetime and its progeny inherit these alterations. For example, if all giraffes stretch their necks during their lifetimes to reach the ever higher foliage, all offspring will inherit this acquired "stretched" neck and begin their lives where their parents have left off. Over time, longer- and longer-necked giraffes would evolve.

   A variational scheme of evolution predicts that not all members of a species contribute equally to future generations. Heritable differences among members of the same species result in some being more "fit" and able to produce more offspring than others. Over time, one type of individual is replaced by another. For example, using an economic metaphor, during the 1990s, there was an explosion of internet and "dot.com" companies. A variation scheme of evolution would have suggested that some of these companies would grow, evolve, and prosper while others, based on intrinsic (heritable) differences (management, capitalization, etc.) would become extinct—a prediction that certainly came true.

2. The three principles are: (1) organisms within a species vary from one another, (2) the variation is heritable, and (3) different types leave different numbers of offspring in future generations.

3. The variation required by Darwin's proposed mechanism of evolution can not exist or be maintained if "blending" or nonrandom segregation occurs. In either case, populations will become homogeneous and variation would be rapidly lost. The "particulate" nature of genes described by Mendel allows for segregation of traits generation after generation. In this way, even currently detrimental (and recessive) alleles can be retested as the environment and circumstances change.

4. A geographical race is a population that is genetically distinguishable from other local populations but is capable of exchanging genes with those other local populations.

A species is a group of organisms that exchanges genes within the group but cannot do so with other groups.

Populations that are geographically separated will diverge from each other as a consequence of a combination of unique mutations, selection, and genetic drift. For populations to diverge enough to become reproductively isolated, spatial separation sufficient to prevent any effective migration is usually necessary.

CHALLENGING PROBLEMS

5. A population will not differentiate from other populations by local inbreeding if

$$\mu \geq 1/N$$

so

$$N \geq 1/\mu$$
$$N \geq 10^5$$

6. The rate of loss of heterozygosity in a closed population is $1/(2N)$ per generation.

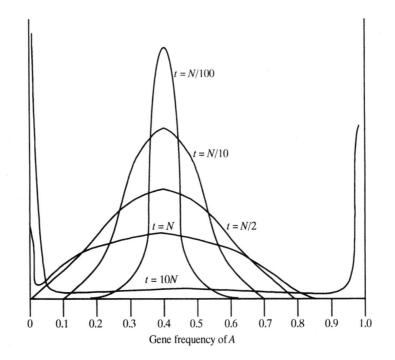

Gene frequency of A

7. A population will not differentiate from other populations by local inbreeding if

   the number of migrant individuals ≥ 1 per generation

   For (a), migration is not sufficient to prevent local inbreeding so the results are roughly the same as seen in problem 6. In the case of (b), there is one migrant per generation so the populations will not differentiate and no locus should be found for which one allele is fixed in some populations and an alternative allele is fixed in others.

8. The mean fitness is calculated by summing the frequency of each progeny class times its fitness. For example, the frequency of $A/A \cdot B/B$ is $p(A)^2 \times p(B)^2$ or $(0.64)(0.81) = 0.52$. This frequency is multiplied by its fitness to give $(0.52 \times 0.95) = 0.49$. Summing for all classes, the mean fitness of population 1 $[p(A) = 0.8, p(B) = 0.9]$ is 0.92. Because selection acts to increase the mean fitness, the frequency of both A and B should increase in the next generation (the $A/A \cdot BB$ class has a fitness of 0.95).

   The mean fitness of population 2 $[p(A) = 0.2, p(B) = 0.2]$ is 0.73. Again, the frequency of both A and B should increase.

   There is a single adaptive peak at $A/A \cdot B/B$. By inspection, the fitness is lowest at $a/a \cdot b/b$ and highest at $A/A \cdot B/B$. The allelic frequency at the peak is 1.0 for both A and B.

9. The mean fitness for population 1 $[p(A) = 0.5, p(B) = 0.5]$ is 0.825. The mean fitness for population 2 $[p(A) = 0.1, p(B) = 0.1]$ is 0.856. There are four adaptive peaks: at $p(A) = 0.0$ or 1.0 and $p(B) = 0.0$ or 1.0. (The mean fitness will be 0.90 and any of these points.) With population 1, the direction of change for both $p(A)$ and $p(B)$ will be random. Both higher or lower frequencies of either allele can result in increased mean fitness (although there are some combinations that would lower fitness). Because population 2 is already near an adaptive peak of $p(A) = 0.0$ and $p(B) = 0.0$, both $p(A)$ and $p(B)$ should decrease to increase the mean fitness.

10. When amino acid changes have been driven by positive adaptive selection, there should be an excess of nonsynonymous changes. The *MC1R* gene (melanocortin 1 receptor) encodes a key protein controlling the amount of melanin in skin and hair. Asian and European populations appear to have experienced positive adaptive selection for more lightly pigmented skin relative to their African counterparts.

11. Both the owl monkey and bush baby are nocturnal while the subterranean mole rat is blind. Genes necessary for colorvision are therefore not necessary for these three species. When species shift their habitats or life styles, mutations that cause inactivation of genes whose functions are no longer necessary is expected.

12. Noncoding sequences. A major constraint on gene evolution is the potential pleiotropic effects of mutations in coding regions. These can be circumvented by mutations in regulatory sequences which play a major role in the evolution of body form. Changes in noncoding sequences provide a mechanism for altering one aspect of gene expression while preserving the role of pleiotropic proteins in other essential developmental processes.

13. Noncoding changes. The difference in pigment expression between spotted and unspotted species of *Drosophila* is more likely due to the differences in cis-acting regulatory sequences that regulate the pigment gene. Mutations in coding regions might be expected to be highly pleiotropic. To test this, the activity of the cis-acting regulatory sequences from the different species could be tested by placing them upstream of a reporter gene and then placing the constructs into *D. melanogaster*. The regulatory sequences from a spotted species should drive high levels of reporter gene expression in just a spot while those from unspotted species should drive low-level expression of the reporter gene across the whole wing blade.

14. Purifying selection, or the "rejection of injurious variations," as Darwin termed it, explains why many protein sequences are unchanged, or nearly so, over vast spans of evolutionary time. A lower-than-expected ratio of nonsynonymous to synonymous changes is a signature of purifying selection. For example, a set of about 500 genes that exist is all Domains-of-life encode proteins whose sequences have been largely conserved over three billion years of evolution. To preserve such sequences, variants that have arisen at random in individuals in tens of millions of species have been rejected by selection over and over again.

15. The text discusses the frequency distribution of haploid chromosome numbers among dicots. Above a chromosome number of 12, even numbers are much more common than odd numbers. This is evidence of frequent polyploidization during plant evolution. For example, if a species of plant with an odd haploid chromosome number undergoes a "doubling" event, the chromosome number becomes even.

16. The α and ß gene families show remarkable amino acid sequence similarities (see table in the companion text). Within each gene family, sequence similarities are greater and in some cases, member genes have identical intron-exon structure.

17. All human populations have high *i*, intermediate I^A, and low I^B frequencies. The variations that do exist among the different geographical populations are most likely due to genetic drift. There is no evidence that selection plays any role regarding these alleles.

18. To test the species distinctness, it is necessary to be able to manipulate and culture *D. pseudoobscura* and *D. persimilis* in captivity. If they cannot be cultured in the laboratory, their species distinctness cannot be established. The

mating behavior compatibility of the different *Drosophila* can be tested by placing a mixture of males of both forms with females of one of the forms to see whether there are any female mating preferences. The same experiment can then be repeated with mixed females and one sort of male and with mixtures of males and females of both forms. From such experiments, patterns of mating preference can be observed. Even if there is some small amount of mating of different forms, this may occur only because of the unnatural conditions in which the test is being carried out.

When matings between different forms do occur, the survivorship of the interpopulation hybrids can be compared with that of the intrapopulation matings. If hybrids survive, their fertility can be tested by attempting to back-cross them to the two different parental strains. As in the case of the mating tests, under the unnatural conditions of the laboratory, some survivorship or fertility of species hybrids is possible even though the isolation in nature is complete. Any clear reduction in observed survivorship or fertility of the hybrids is strong presumptive evidence that they belong to different species.

19.

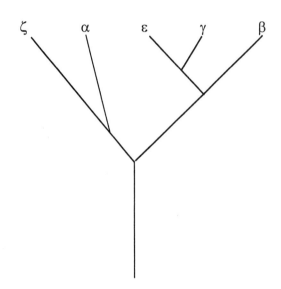

20. For polymorphic sites within a species, let nonsynonymous = a and synonymous = b. For polymorphic sites between the species, let nonsynonymous = c and synonymous = d. If divergence is due to neutral evolution, then

$$a/b = c/d$$

If divergence is due to selection, then

$$a/b < c/d$$

However, in this example, $a/b = 20/50 > c/d = 2/18$, which fits neither expectation.

Because the ratio of nonsynonymous to synonymous polymorphisms ($^a/_b$) is relatively high, the gene being studied may encode a protein tolerant of substitution (like fibrinopeptides, discussed in the companion text). The relatively fewer species differences may suggest that speciation was a recent event, so few polymorphisms have been fixed in one species that are not variants in the other.

20

Gene Isolation and Manipulation

BASIC PROBLEMS

1. **a.** The following are examples of hybridization of single-stranded DNAs discussed in this chapter: use of primers (DNA sequencing, PCR), hybridization of sticky ends in cloning, probe hybridization (Southern blotting, Northern blotting, diagnosis of mutations, identification of clones, and *in situ* hybridization).

    b. The following are examples of proteins that bind to and act on DNA discussed in this chapter: restriction enzymes, DNA polymerase (*Taq* polymerase), reverse transcriptase, and DNA ligase.

2. Reverse transcriptase polymerizes DNA using RNA as a template. This RNA:DNA hybrid is then treated with sodium hydroxide to degrade the RNA. (NaOH hydrolyzes RNA by catalyzing the formation of a 2′-3′cyclic phosphate.)

3. **a.** Recombinant DNA is the term used to describe techniques used to generate hybrid molecules consisting of DNA from two different sources. In this use, recombinant refers to the artificial joining of DNA from different sources.

    b. Recombinant frequency is used to describe the statistical likelihood that two loci in the same organism will assort together or separately during meiosis. Here, recombinant refers to the combinations of loci that are different from the parent.

4. Ligase is an essential enzyme within all cells that seals breaks in the phosphate-sugar backbone of DNA. During DNA replication it joins Okazaki fragments to create a continuous strand, and in cloning, it is used to join the various DNA fragments with the vector. If it was not added, the vector and cloned DNA would simply fall apart.

5. Each cycle takes five minutes and doubles the DNA. In one hour there would be 12 cycles so the DNA would be amplified $2^{12} = 4096$-fold amplification.

6. The great advantage of PCR is that fewer procedures are necessary compared with cloning. However, it requires that the sequence be known, and that the primers are each present once in the genome and are sufficiently close (less than 2 kb). If these conditions are met, the primers determine the specificity of the DNA segment amplified and would be most efficient.

7. Resequencing the relevant gene should be done as that will tell you if the original sequence was correct. You can then use the mutant sequence in a gene replacement experiment (depending on the organism) to see if the mutant phenotype is actually the result of the sequence variation you detected.

8. You could isolate DNA from the suspected transgenic plant and probe for the presence of the transgene by Southern hybridization.

9. Follow the protocol discussed in Figure 11-38 in the companion text. Basically, inject the rat growth hormone gene (RGH) into fertilized eggs of *lit/lit* mice. These eggs are then implanted into a surrogate mother and any resulting offspring mated to test their offspring to see if any are transgenic for RGH. Because RGH will be inherited in a dominant fashion, any large and transgenic offspring will be heterozygous at this point. The transgenic siblings will need to be mated to each other in order to generate mice that are homozygous for the RGH transgene.

10. The commercial cloning of insulin was into bacteria. Bacteria are not capable of processing introns. Genomic DNA would include the introns, while cDNA is a copy of processed (and thus intron-free) mRNA.

11. a. Because the actin protein sequence is known, a probe could be synthesized by "guessing" the DNA sequence based on the amino acid sequence. (This works best if there is a region of amino acids that can be coded with minimal redundancy.) Alternatively, the gene for actin cloned in another species can be used as a probe to find the homologous gene in *Drosophila*. If an expression vector was used, it might also be possible to detect a clone coding for actin by screening with actin antibodies.

    b. Hybridization using the specific tRNA as a probe could identify a clone coding for itself.

12. a. The transformed phenotype would map to the same locus. If gene replacement occurred by a double crossing-over event, the transformed cells would not contain vector DNA. If a single crossing-over took place, the entire vector would now be part of the linear *Neurospora* chromosome.

b. The transformed phenotype would map to a different locus than that of the auxotroph if the transforming gene was inserted ectopically (i.e., at another location).

13. Size, translocations between known chromosomes, and hybridization to probes of known location can all be useful in identifying which band on a PFGE gel corresponds to a particular chromosome.

14. Conservatively, the amount of DNA necessary to encode this protein of 445 amino acids is $445 \times 3 = 1335$ base pairs. When compared with the actual amount of DNA used, 60 kb, the gene appears to be roughly 45 times larger than necessary. This "extra" DNA mostly represents the introns that must be correctly spliced out of the primary transcript during RNA processing for correct translation. (There are also comparatively very small amounts of both 5´ and 3´ untranslated regions of the final mRNA that are necessary for correct translation encoded by this 60-kb of DNA.)

15. The typical procedure is to "knock out" the gene in question and then see if there is any observable phenotype. One methodology to do this is described in in the companion text. A recombinant vector carrying a selectable gene within the gene of interest is used to transform yeast cells. Grown under appropriate conditions, yeast that have incorporated the marker gene will be selected. Many of these will have the gene of interest disrupted by the selectable gene. The phenotype of these cells would then be assessed to determine gene function.

CHALLENGING PROBLEMS

16. **a.** To ensure that a colony is not, in fact, a prototrophic contaminant, the prototrophic line should be sensitive to a drug to which the recipient is resistant. A simple additional marker would also achieve the same end.

b. Use a nonrevertible auxotroph as the recipient (such as one containing a deletion.)

17. a.

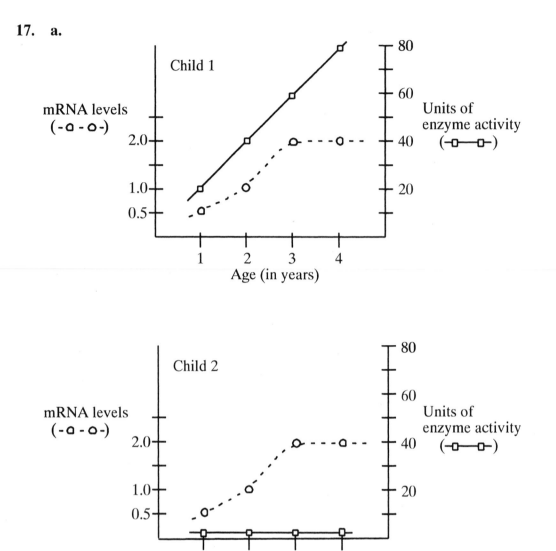

b. and c. Very low levels of active enzyme are caused by the introduction of an *Xho* site within the D gene. Because the size of the nonfunctional enzyme has not changed, the most likely event was a point mutation within the coding region of the gene that also created a new *Xho* site. It is likely that this point mutation also altered the active site destroying enzyme activity.

d. Individual 1 would be defined as homozygous normal, while individual 2 is homozygous mutant. If either were heterozygous, there would be three bands hybridizing to the probe on the Southern blot.

18. a. The gel can be read from the bottom to the top in a 5´-to-3´ direction. The sequence is

$$5´ \; \text{TTCGAAAGGTGACCCCTGGACCTTTAGA} \; 3´$$

b. By complementarity, the template was

$$3'\ A\,A\,G\,C\,T\,T\,T\,C\,C\,A\,C\,T\,G\,G\,G\,G\,A\,C\,C\,T\,G\,G\,A\,A\,A\,T\,C\,T\ 5'$$

c. The double helix is

$$5'\ T\,T\,C\,G\,A\,A\,A\,G\,G\,T\,G\,A\,C\,C\,C\,C\,T\,G\,G\,A\,C\,C\,T\,T\,T\,A\,G\,A\ 3'$$
$$3'\ A\,A\,G\,C\,T\,T\,T\,C\,C\,A\,C\,T\,G\,G\,G\,G\,A\,C\,C\,T\,G\,G\,A\,A\,A\,T\,C\,T\ 5'$$

d. Open reading frames have no stop codons. There are three frames for each strand, for a total of six possible reading frames. For the strand read from the gel, the transcript would be

$$5'\ U\,C\,\mathbf{\underline{U\,A\,A}}\,A\,G\,G\,U\,C\,C\,A\,G\,G\,G\,G\,U\,C\,A\,C\,C\,U\,U\,U\,C\,G\,A\,A\ 3'$$

And for the template strand

$$5'\ U\,U\,C\,G\,A\,A\,A\,G\,G\,\mathbf{\underline{U\,G\,A}}\,C\,C\,C\,C\,U\,G\,G\,A\,C\,C\,U\,U\,\mathbf{\underline{U\,A\,G}}\,A\ 3'$$

Stop codons are in bold and underlined. There are a total of four open reading frames of the six possible.

19. The region of DNA that encodes tyrosinase in "normal" mouse genomic DNA contains two *Eco*RI sites. Thus, after *Eco*RI digestion, three different-sized fragments hybridize to the cDNA clone. When genomic DNA from certain albino mice is subjected to similar analysis, there are no DNA fragments that contain complementary sequences to the same cDNA. This indicates that these mice lack the ability to produce tyrosinase because the DNA that encodes the enzyme must be deleted.

20. Plant 1 shows the typical inheritance for a dominant gene that is heterozygous. Assuming kanamycin resistance is dominant to kanamycin sensitivity, the cross can be outlined as follows:

$$kan^R/kan^S \times kan^S/kan^S$$
$$\downarrow$$
$$1/2\ kan^R/kan^S$$
$$1/2\ kan^S/kan^S$$

This would suggest that the gene of interest would be inserted once into the genome.

Plant 2 shows a 3:1 ratio in the progeny of the backcross. This suggests that there have been two unlinked insertions of the *kan^R* gene and presumably the gene of interest as well.

$$kan^{R1}/kan^{S1} \, ; kan^{R2}/kan^{S2} \, \times \, kan^{S1}/kan^{S1} \, ; kan^{S2}/kan^{S2}$$

$$\downarrow$$

$1/4 \quad kan^{R1}/kan^{S1} \, ; kan^{R2}/kan^{S2}$

$1/4 \quad kan^{R1}/kan^{S1} \, ; kan^{S2}/kan^{S2}$

$1/4 \quad kan^{S1}/kan^{S1} \, ; kan^{R2}/kan^{S2}$

$1/4 \quad kan^{S1}/kan^{S1} \, ; kan^{S2}/kan^{S2}$

21. Assuming that the DNA from this region is cloned, it could be used as a probe to detect this RFLP on Southern blots. DNA from individuals within this pedigree would be isolated (typically from blood samples containing white blood cells) and restricted with *Eco*RI, and Southern blots would be performed. Individuals with this mutant CF allele would have one band that would be larger (owing to the missing *Eco*RI site) when compared with wild type. Individuals that inherited this larger *Eco*RI fragment would, at minimum, be carriers for cystic fibrosis. In the specific case discussed in this problem, a woman that is heterozygous for this specific allele marries a man that is heterozygous for a different mutated CF allele. Just knowing that both are heterozygous, it is possible to predict that there is a 25 percent chance of their child's having CF. However, because the mother's allele is detectable on a Southern blot, it would be possible to test whether the fetus inherited this allele. DNA from the fetus (through either CVS or amniocentesis) could be isolated and tested for this specific *Eco*RI fragment. If the fetus did not inherit this allele, there would be a 0 percent chance of its having CF. On the other hand, if the fetus inherited this allele, there would be a 50 percent chance the child will have CF.

22. The promoter and control regions of the plant gene of interest must be cloned and joined in the correct orientation with the glucuronidase gene. This places the reporter gene under the same transcriptional control as the gene of interest. The companion text discusses the methodology used to create transgenic plants. Transform plant cells with the reporter gene construct, and as discussed in the text, grow into transgenic plants. The glucuronidase gene will now be expressed in the same developmental pattern as the gene of interest and its expression can easily be monitored by bathing the plant in an X-Gluc solution and assaying for the blue reaction product.

23. **a. and b.** During Ti plasmid transformation, the kanamycin gene will insert randomly into the plant chromosomes. Colony A, when selfed, has $3/4$ kanamycin-resistant progeny, and colony B, when selfed, has $15/16$ kanamycin-resistant progeny. This suggests that there was a single insertion into one chromosome in colony A and two independent insertions on separate chromosomes in colony B. This can be schematically represented by showing a single insertion within one of the pair of chromosome "A" for colony A.

Chromosomes "A"

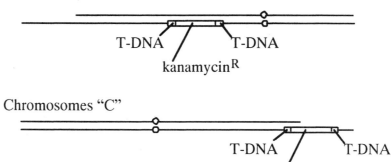

and two independent insertions into one of each of the pairs of chromosomes "B" and "C" for colony B.

Chromosomes "B"

Chromosomes "C"

Genetically, this can be represented as

Colony A $\qquad kan^{R_A}/kan^{S_A}$
Colony B $\qquad kan^{R_B}/kan^{S_B}$; kan^{R_C}/kan^{S_C}

When these are selfed $\qquad kan^{R_A}/kan^{S_A} \times kan^{R_A}/kan^{S_A}$

$\quad 1/4 \qquad kan^{R_A}/kan^{R_A}$
$\quad 1/2 \qquad kan^{R_A}/kan^{S_A}$
$\quad 1/4 \qquad kan^{S_A}/kan^{S_A}$

kan^{R_B}/kan^{S_B} ; $kan^{R_C}/kan^{S_C} \times kan^{R_B}/kan^{S_B}$; kan^{R_C}/kan^{S_C}

$\quad 9/16 \qquad kan^{R_B}/-$; $kan^{R_C}/-$
$\quad 3/16 \qquad kan^{R_B}/-$; kan^{S_C}/kan^{S_C}
$\quad 3/16 \qquad kan^{S_B}/kan^{S_B}$; $kan^{R_C}/-$
$\quad 1/16 \qquad kan^{S_B}/kan^{S_B}$; kan^{S_C}/kan^{S_C}

EXPLORING GENOMES

Table of Contents

Introduction

Computer science has had a major impact on the biological sciences, and this is particularly true in the field of sequence analysis. As sequencing technology improved and became highly automated in the 1990s, researchers around the world accumulated a wealth of information that rapidly grew beyond the scope of what scientists could analyze independently. Scientists and government officials with considerable foresight lobbied for centralized institutions that could store this information and make these resources available to researchers worldwide through the internet. At the same time, programmers were developing potent analysis tools that could mine the information in the databases for comparisons at a very detailed level. With the advent of easy and near universal internet web access, researchers have come to rely on these institutions to an ever increasing extent. Genomics, the study of whole genomes, and bioinformatics, the development and use of computer tools to analyze them, now contribute to virtually all areas of biological science. It is safe to say that understanding how to make use of these resources is an essential skill for anyone in the fields of biology and biomedical sciences.

The resources available are located on thousands of different web sites and are quite varied in size and nature. Some are very narrowly focused, such as databases created for a particular organism or sequencing project. Alternatively, a site might be set up simply to allow the use of a particular type of analysis or a new software tool. Some sites may represent the efforts of a single laboratory. Others, such as the National Center for Biotechnology Information (NCBI) in Washington, D. C., have as their mandate the collection of all publicly available sequence and related information and the development of software tools for its use.

The curators and computer programmers at comprehensive centers such as NCBI compile sequence and related data as it becomes available from sequencing centers around the world and present it in a form that is easily accessible to researchers. Their efforts are coordinated with similar genome centers in many countries. NCBI and its counterparts, such as the European Molecular Biology Laboratory (EMBL) in Germany, and the DNA Database of Japan (DDBJ), are the primary centers for the collection and archiving of sequence data.

The software tools that are used for access and analysis differ at the various web sites. First and foremost, an archival site such as NCBI allows researchers to retrieve the original sequence record and all subsequent updates and modifications for a gene, protein, or genome. The sites also provide tools for gene alignment that are essential in answering questions regarding the conservation of genes among organisms. This often allows information from major genetic systems such as yeast, Drosophilia, or Arabidopsis to provide clues as to the biological function of a human gene, or vice versa. At another level, the programmers actively participate in designing new software tools for problems as diverse as identifying the coding regions within newly sequenced genomes, for linking gene discoveries to human disease, and for examining and comparing the three-dimensional structures of proteins.

Students trying to find their way through one of these web sites are often intimidated by the unfamiliarity of the material and terms, as well as a sense that their computer skills are not as advanced as they would like. They needn't worry on either count. Web-based access at its best is designed to allow very sophisticated analyses to be carried out with little or no detailed knowledge of the underlying programming. All the student needs is a little guidance.

The tutorials in this book and on its accompanying web site at **www.whfreeman.com/young** focus almost entirely on a single site, the National Center for Biotechnology Information. It is one of the world's largest resources and it has a wide range of useful tools and presentations available. The tutorials here are designed to introduce the basics of genomic analysis at a level appropriate for college students. These tutorials give students initial guidance and practice in exploring the breadth and depth of the resources by helping them understand the structure and content of the databases, and facilitating direct practice in the use of search and alignment tools.

Each tutorial is written as an interactive guide to an aspect of the NCBI site. The user's browser window is split into two parts. Guidance is provided as text on one side of the screen, while the main window is directly connected to NCBI. This latter window runs live on the NCBI computers. Using real examples from the research literature, the tutorial walks students through the exercise and tells them precisely how to run the programs and interpret NCBI results.

These tutorials are not intended to be comprehensive. In many ways, they simply scratch the surface of the resources. The do, however, allow a student to gain practical skills and to do independent research very quickly. A student that becomes familiar with this material will have no problem going forward to more advanced use of the resources. NCBI itself has a broad array of tutorials associated with their web site. Many of these are at a more advanced level and the introductory ones provided here serve as a bridge to them.

The tutorials are designed to display some of the major public resources of the NCBI, keeping in mind the level and interests of beginning genetics student. For each tutorial the searches are run with known genes or sequences and for the most part examples from our major genetic systems such as *E. coli, Drosophila,* yeast, *Arabidopsis,* and human. Usually a fairly common protein or gene is chosen so that it might be familiar to a student in genetics. The choice of example is sometimes dictated by a desire for a relatively simple result (not always possible).

The tutorials include and overview of the NCBI site as well as examples of how to use the information resources such as Entrez, PubMed, OMIM, and the Cancer Genome Anatomy Project. There are several tutorials dealing with gene alignments using BLAST, as well as searches for conserved domains using pattern and profile searching. The student can explore the 3D structure of nucleic acids and proteins and even search the database for proteins with similar 3D structure. Lastly, taxonomy resources are examined and a multiple alignment of a gene family is used (ClustalW) in order to draw a phylogenetic tree. In all cases, students are asked to go back to the program using an input of their choice (with some suggestions made) so that they can reinforce their knowledge.

Database resources such as NCBI and the European Molecular Biology Laboratory (EMBL) are dynamic and constantly being improved. They increase in size daily and frequently add to the services that they offer. The inevitable consequence is that occasionally hyperlinks are altered. Although these tutorials are frequently monitored to ensure that they work, sometimes a glitch occurs. We would appreciate hearing of any problems so that they can be corrected. Please email us at **techsupport@bfwpub.com.**

Detailed Table of Contents

inheritance pattern for conditions such as Huntington disease. This will also allow you to explore the nature of the gene and the various sequence resources available for it.

7. Finding Conserved Domains

Conserved protein sequences are a reflection of conservation of amino acid residues necessary for structure, regulation, or catalytic function. Oftentimes, groups of residues can be identified as being a pattern or signature of a particular type of enzyme or regulatory domain. You will learn how to find the conserved domains in a complex protein by searching databases of conserved patterns. In addition, this will allow you to examine the domain structure of proteins and the way in which this is conserved through evolution.

8. Clusters of Orthologous Groups

As the sequence databases grow, we increasingly find similar genes in different species. These orthologues can be analyzed to investigate their degree of conservation and their distribution in the phylogenetic tree. In the Genomics tutorial at the Web site www.whfreeman.com/iga9e, you will learn how to perform such investigations using the COGs database, which contains information on the conservation and distribution of orthologues gleaned from fully sequenced genomes.

9. Determining Protein Structure

Protein function depends on 3D structure, which is in turn dependent on the primary crystallography or NMR. Databases at NCBI store the 3D structure for all nucleic acids and proteins where information is available. In the absence of direct experimental information, powerful programs can be used to compare the 3D structure of one protein to another and even to search the database based on a direct structural comparison. We'll try one in this tutorial.

10. Functional Analysis

We now have many fully sequenced genomes to play with. Their availability allows computer analysis like that we saw used to analyze the COGs database. It also allows us to design experiments to test which genes function in which processes and how the various gene products interact with each other. In the Genomics tutorial at the Web site www.whfreeman.com/iga9e, you will see how to investigate these questions at the whole genome level using techniques such as gene deletion to examine the loss of function phenotype or methods to investigate protein-protein interaction on a large scale.

11. Exploring the Cancer Genome Anatomy Project

The Cancer Anatomy project at NCBI is a specialized database focused on a particular subset of data. It brings together, in a graphical and searchable format, all known information about individual genes, map positions, and chromosomal location for genes associated with cancer. We will explore gene expression data, gene mutations, and chromosomal aberrations associated with the various tumor types.

12. Measuring Phylogenetic Distance

The NCBI database now has at least some sequence data from over 100,000 different organisms. These data can be used to build a phylogenetic tree and a common taxonomy for all organisms. The taxonomy database at NCBI has exploited this information (and other resources) and allows us to examine the various entries organized by organism and their position in the phylogenetic tree. Sequence data allows us to quantitatively estimate the evolutionary distance among organisms based on the extent of sequence divergence. We will examine this using a gene that is highly conserved in all organisms. By doing a multiple alignment of the sequences using the Clustal and Phylip programs, we will be able to estimate the evolutionary distance among the various organisms and to draw a simple phylogenetic tree.

13. Environmental Genomics

As we have seen, high capacity DNA sequencing centers provide us with the sequence of whole genomes. As you will see in this tutorial at the Web site www.whfreeman.com/iga9e they also can be used to investigate the microbial diversity in our environment. Much to the amazement of the biological community, the analysis of DNA extracted from the environment shows that the vast majority of microbes and microbial biomass have never been cultured or previously identified. Nevertheless, using the techniques of high capacity sequencing coupled with gene comparison and alignment, we are able to piece together genes and even genomes from creatures we have never seen.

Introduction to Genomic Databases

Starting link: http://www.ncbi.nlm.nih.gov

Everyone has a difficult time keeping up with the flow of new information. This is particularly true in biology now as the pace of discovery accelerates. Databases have become an essential tool for accumulating and archiving raw data. They also play a major role in analyzing and presenting information to researchers and the public in an easily accessible form. In this *Exploring Genomes* tutorial we will survey one of these resources: The National Center for Biotechnology Information (NCBI) located in Washington, D.C.

One of the roles of NCBI is to archive raw DNA sequence data. The sequence information comes from research efforts in laboratories around the world as well as from large-scale, dedicated genome sequencing centers. The resulting database is referred to as GenBank. GenBank shares its resources with the European and Japanese equivalents so that there are three primary public repositories of such information in the world. Because of the automation of DNA sequencing over the past decade, these databases are increasing in size exponentially. GenBank includes the sequences of the *E. coli*, *Drosophila*, and human genome, as well as data from thousands of other species. At present it comprises millions of DNA sequences with a cumulative length of over 100 billion base pairs. You can see the rate of growth by clicking here.

[http://www.ncbi.nlm.nih.gov/Genbank/genbankstats.html]

Growth of GenBank

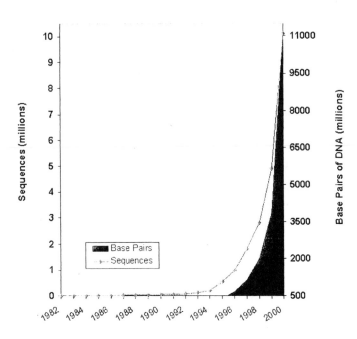

Now let's go back to the NCBI home page, the public access point to its many resources (http://www.ncbi.nlm.nih.gov/). A quick glance at this page shows that NCBI contains far more than just the sequence repository. It is a rich source of information on all aspects of genetics and genomics. All of the divisions are searchable and information ranging from gene sequences, to the position of a locus on a human chromosome, to direct access to the scientific literature dealing with a particular gene, is immediately accessible.

There are seven major categories of service listed across the black navigation bar at the top of the NCBI homepage. These are PubMed, All Databases, BLAST, OMIM, Tax Browser, and Structure. In addition there are specialized projects and databases listed on the right hand side of the page. In this *Exploring Genomes* tutorial we will take a quick look at some of these resources. Subsequent tutorials will explore a particular resource in much greater depth. Let's click on each of the six resources at the top in turn. Click on **'PubMed'** first.

PubMed is the NCBI gateway to the biomedical research literature. It is a searchable database and information can be retrieved based on combinations of parameters such as author, subject key words, or organism. A complex query can be entered and a list of publications matching it will be returned. For instance, we could enter a simple search by author. If we entered **Hartwell LH** (one of the winners of the 2001 Nobel Prize in

Medicine) and pressed **'Go,'** PubMed would return a list of his current publications. Give it a try.

___

The list of Hartwell's publications is 5 pages long. Only the top and therefore most recent papers are displayed on the first page. Each paper is linked to its abstract and sometimes the full text of the articles. We will explore PubMed much more fully in a later tutorial. Let's go back and try another NCBI division by clicking the **NCBI icon** on the top left to get to the NCBI homepage again, and then clicking the button for **'All Databases.'**

___

This button opens a search engine called 'Entrez' that links all of the NCBI databases together. This allows access to everything from sequence to structure. The options are listed. The PubMed link that we looked at initially was only one of these options. Let's try searching for the human keratin protein sequence in the Protein database. Choose **'Protein'** from the menu, and then type **'keratin AND human'** in the text box. Now press **'Go.'**

___

The search returns a list of database entries including keratin-associated proteins and keratins themselves. Note that only the first 20 entries of a total of 900 are displayed on the first page! For each entry, clicking on the associated links will display the sequence, related information, and links to other parts of the databases. We can explore these later. Let's go back to the **NCBI** homepage and try the **'BLAST'** button.

___

BLAST is a powerful nucleic acid or protein alignment tool. It allows us to dynamically search the sequence databases (all 100 billion base pairs!) to find similar sequences in different organisms. It is extremely versatile and comes in many different forms for doing different types of searches. The underlying method is the same in each case, however. This is our standard software tool for doing such searches. It is very important and we will devote an entire tutorial to its use later. For now, let's move back to the **NCBI** homepage and click on to **'OMIM.'**

___

OMIM is the Online Mendelian Inheritance in Man database. Note from the overview on the OMIM homepage below that it integrates the known Mendelian genetics of human disease with the resources made available through Entrez at NCBI. We will explore this in detail later but for now let's open the **OMIM Statistics** link, under OMIM Facts on the left-hand menu.

___

The OMIM statistics page gives an overview of the breadth of the resource. Note the categorization by Mendelian inheritance pattern. Note also that over 17,000 entries are available with at least some information. Keep in mind that we believe there are about 25,000 human genes! All genes will not necessarily be associated with a disease or visible phenotype and therefore for many genes little information is yet available. The Human Genome Project has provided a glimpse at large numbers of new genes of unknown function, reminding us of how much remains to be done.

Next, click back to the **NCBI** homepage and on to the **Taxonomy** database.

The Taxonomy section groups all data by taxonomic classification. You may type in a species name to find out if any sequence information is available; for the major genetic systems, simply click on the links on the Taxonomy Browser home page. Try *Caenorhabditis elegans*, a nematode worm, one of our most powerful model genetic systems whose full genome sequence was recently determined.

This organism-specific page gives important information regarding phylogenetic lineage as well as the number of sequences of various types that have been deposited. These groupings may be called up at will and each of the genes listed are linked to the Entrez system. One important use of the groupings is to restrict other types of searches. For instance a BLAST search can be launched from the BLAST page and the database searched restricted to *Caenorhabditis elegans* only (or any other species).

Now, let's go to the last major subdivision, **'Structure.'**

The structure database contains the 3D structure for all nucleic acids and proteins whose shape has been determined by X-ray crystallography or nuclear magnetic resonance. We can call up these 3D models at will. The structure database is associated with the VAST program that allows for 3D structural comparisons among different proteins. It also will search the database on the basis of structure. These are very powerful tools and we will give them a try later in the term.

Now, let's go back to **NCBI** home page to see some of its other resources.

Apart from the basic databases and access software, NCBI has a wide variety of highly specialized databases and analyses. These focus on particular problems or interest groups. Some are listed on the right hand side of the NCBI homepage below. We will just touch on a couple of them for now to get a sense of their capabilities.

The first is the Map Viewer. This allows us to visualize the various chromosomes and the genetic loci on them. Let's take a look by clicking on **'Map Viewer'** under the Hot Spots list. From the resulting screen choose **'Homo sapiens (human)'** from the Mammals category.

The human genome view is a visualization of the full chromosome set. Clicking on any chromosome number will expand the view of that particular chromosome. Let's try the Y chromosome.

The graphic of the Y chromosome is expanded so that we can see its banding pattern as seen by a cytologist. Aligned along it are the blocks of DNA that have been sequenced and represented here (contigs) followed by the genes located on these pieces of DNA. The genes are represented in the first column as the compiled summary data in Unigene (another of the NCBI databases). Subsequent columns provide links to the various genes that have been annotated on the DNA sequence. Let's look more closely at one by clicking on any gene in the Unigene list, then on the name in the pop-up. This will open a small window. Click on the link.

The Unigene summary is an annotated view. It is compiled by a curator and takes into account all known information regarding the sequence. Unigene also includes links to OMIM as well as a comparison to the most similar genes in other organisms. It is a very rich source of information. All of the data is cross-referenced by links into Entrez and various other databases that provide the gene sequence and research literature references. Ultimately the entire human genome will be available in finished form from this database.

Now let's go back to the **NCBI** homepage and click on the first resource under Hot Spots, **NCI-CGAP**.

The Cancer Genome Anatomy Project is the last topic we'll briefly explore. This database focuses only on tumor tissue and strives to provide cross-referenced information for all genes thought to be involved in cancer. Each of the subsections enters the database with a different type of approach. Ultimately you can explore the database to great depth, searching out the genes involved, chromosomal locations and aberrations, and biochemical pathways. All of this information is related to the tissues and cells where the tumor that you are interested in originates.

Over the course of these *Exploring Genomes* tutorials, we will look at parts of these databases in much greater detail. A facility for handling them is an essential skill for a modern biologist. Ultimately, the only way to familiarize yourself with a resource of this type is to go to the web site (http://www.ncbi.nlm.nih.gov/) and start exploring some of the links. You might start by searching for the answer to the following simple questions:

- What is the publication history of your Biology professors and in particular your Genetics instructors?
- What organisms do they work with?
- What types of question are they trying to answer?

Learning to use Entrez

Starting Link: http://www.ncbi.nlm.nih.gov

The Entrez retrieval system is the entry point for searching most of the material at NCBI. Its strength is that it provides links between related types of information. For instance, in storing a DNA sequence file, the file is associated with the protein translation of the sequence, with the literature reference, and with links to similar genes or proteins in other organisms. It also provides a characterization of any notable features, such as conserved regions, and ultimately chromosomal location and 3D structure of the gene product. The retrieval system moves relatively effortlessly among these various types of data. The links are updated as new data are added to the databases (or new databases are developed) and the resource becomes richer and richer over time.

We looked very briefly at Entrez in the Introductory tutorial. Now let's learn to use it bywalking through an analysis of the dystrophin gene in humans. As you know from Chapter Two of your textbook, this is the gene responsible for causing Duchenne muscular dystrophy (DMD).

Click on **'All Databases'** on the top blue bar to access the Entrez home page from the NCBI home page.

All of these databases are ultimately linked within the framework of the Entrez retrieval system.

Let's start with a research literature search for dystrophin on PubMed. Since it is a well-characterized protein involved in an important human disease, expect a long list of retrievals.

Choose **'PubMed'** from the Search drop-down menu, type **'dystrophin'** into the text box, and then click **'Go.'**

The search returns a rather daunting several thousand entries, obviously a very active research area. They list in reverse chronological order and only the last few are displayed here. Notice the titles of some of the papers. Clicking on the authors' names will link you to the publication information and abstracts. Note that the reference numbers change as new publications are added to the database.

For this exercise, we are going to take a look at a paper written in 1987, when the molecular genetics of the dystrophin locus and the Mendelian genetics of Duchenne muscular dystrophy were converging.

Let's call up the Hoffman, Brown and Kunkel (1987) reference. Since it is so old, and the articles are listed in reverse chronological order, it will take a lot of clicking through the pages to get to it. A faster way to reach the article is to do a search with limits. Click **'Limits'** on the light blue bar above the Display button. In the Published in the Last field, choose 'specific date,' then type in **'1987 01 01.'** In the field **To,** type in **'1987 12 31.'** Now press **'Go.'**

When the search is completed, you should see the paper by Hoffman, Brown and Kunkel (1987) near the top of your screen. Now click on the **authors' names** to link to the information related to this paper.

The abstract of the paper is displayed as well as a link to the full text article. For most articles there is a registration and charge to go beyond this point. Also included is a set of links on the right (in blue). These point to closely related research articles, to the gene and inheritance pattern, as well as books and other resources. Clearly we could pursue an extensive library research project related to this gene simply by following a number of these links. For the moment we will find the chromosome map of the gene responsible for DMD.

From this page, click on **'OMIM'** by opening the light blue 'Links' near the title of the paper or go to the Entrez OMIM link, type **'dystrophin'** and **'Go.'**

This link will connect us to all the literature related to DMD and dystrophin. For now, though, we just want to see the gene map locus. Click on **'Xp21.2, 12q21.'**

Now you should see a chart with all the gene locations related to DMD and dystrophin. Click on the first location, **'Xp21.2.'**

This will take us to Entrez Genome page.

The graphic that appears is a view of the human X-chromosome. The DMD (Duchenne Muscular Dystrophy) gene is highlighted. Clicking on the DMD link itself will take us to the LocusLink page where there is an extensive compilation of information about the gene and its expression. Click on **'DMD'** and a window will open to a DMD link.

LocusLink is a very rich resource of data (scroll down through the pages and examine it) that links back to the other databases at NCBI. By now you should be getting a sense of the extent of the interconnection built into the Entrez structure. Clearly we could pursue this gene in many directions from this page. However, let's go back to the Entrez home page and initiate our search for dystrophin by another route.

Click **'All Databases'** on the black menu bar.

From the Entrez page we could start our search a different way with a search for the Protein record for dystrophin. Choose **'Protein'** from the menu, type **'dystrophin'** into the text box, and then press **'Go.'**

The output is very long with hundreds of entries, including various alleles of dystrophin and many dystrophin-related proteins. The sequence we are interested in has the identifier P11532. This identifier number is the **accession number** and was permanently attached to the sequence file when it was first archived in the database.

You can find P11532 by clicking through the pages of results until you see it, or you can do another search in the Protein database for **'P11532.'**

When you find the sequence, you will see that on its right are a number of links. These include BLink, which compares the sequence to others in the database, a link to the nucleotide sequence, one to related sequences, and one to the taxonomy database. Keep in mind that many of these links are displaying different aspects of the NCBI information in different ways. The underlying databases are the same. Let's click on **'BLink.'**

The BLink program displays a graphic of an alignment of our gene with all related sequences in the database. Note that there are many dystrophin entries and the graphic on the left shows the regions of similarity in sequence. Some of these are different isoforms and alleles; some are from other species. Some are older records. NCBI is an archival database, and once generated, all records are kept, although they can be updated. All of these records are available from this page simply by clicking the links. A researcher would use this resource to examine the various records for the gene, and to see, in a simple graphical way, different forms of the protein in humans, or the extent to which dystrophin from other species is similar to the human form.

From here, let's look at the dystrophin protein record itself by clicking on the link **'NP_003997'** in the Accession number column.

The page that comes up is the protein record for our accession number. This record is a computer translation of the DNA sequence determined by sequencing the mRNA for the protein. Computers require standard formats to store data, and humans need accessible formats to view it. This file (in a human readable form) is a **GenBank flatfile.** The display you see is the standard output format for GenBank. It has a specific list of sections, each with a particular type of information. It also includes highlighted links to other aspects of the databases. We don't have to worry about them all, but let's look at a few.

The locus is identified by the accession number, at the top of the record. The DEFINITION line shows it as "dystrophin, muscle – human."

Scroll down to see the REFERENCE section, which gives us all of the literature references related to the sequencing and characterization of this protein.

Further down, the FEATURES list tells us about various predicted or biochemically verified conserved domains, regulatory sites, and binding sites for other proteins. The features are arranged in order starting at the amino terminus of the protein (designated 1) and their location is identified by the amino acid position along this very large protein (3685 amino acids).

Lastly, at the bottom of the record is the protein sequence itself in single letter code.

The format of the GenBank flatfile is a rich source of information and you should familiarize yourself with it. There are other file formats, however, that are much simpler and sometimes more appropriate to use, particularly if all you want is the sequence for insertion into another program. Some analytical programs will accept the accession number as input, but many require that you cut and paste in the sequence. One of the simplest and most widely used formats is the **FASTA** format, which was developed for one of our important search and alignment programs (FASTA).

We can see the FASTA version of our dystrophin protein by toggling the display drop-down menu (at the top left side of the page) to **'FASTA'** and pushing the 'Display' button.

The format has one line of information beginning with the > sign and followed by the accession number and species information. Starting on the next line is the sequence of the dystrophin protein. Notice that there are no line numbers or punctuation—it is perfect for cutting and pasting.

Now let's go back and learn how to execute more specific searches in Entrez. Close the window that we've been working in, and then click **here** to get back to the Entrez home page. [http://www.ncbi.nlm.nih.gov/Entrez/]

Oftentimes in searching for information on a protein such as dystrophin, we can be overwhelmed with the hundreds or thousands of responses. If you have a more narrow interest it is possible to phrase searches using Boolean operators (AND, OR, NOT). For example if we searched the protein database for "dystrophin AND chicken" then our output is sharply focused. We only get a small number of entries including dystrophin and some dystrophin-related proteins from chickens.

See how many results you get when you select the **Protein** database, write **'dystrophin AND chicken'** in the text box, and then press **'Go.'** (Note: The Boolean operators must be written in ALL CAPS.)

There are other ways to limit searches as well. For example, since there are so many human records for such an important human genetic disease gene locus, we might want to exclude human data from the list.

Search the protein database for **"dystrophin NOT human"** to simplify the output.

Now we get everything but human and the list shrinks from thousands to a few hundred entries.

There is really only one way to become conversant with such an important resource. Go to the website at NCBI (http://www.ncbi.nlm.nih.gov/) and continue to explore the Entrez program. This is one of the most useful tools for any biologist and could be considered an essential skill.

You have already examined the publications produced by your Genetics professor. Choose one of those publications related to a particular gene, protein, or organism and search through Entrez to find the associated links. Start your search from the publication you've chosen in PubMed. Using the links on the records, go to Entrez and input a gene or protein name that you have gleaned from one of the abstracts of the papers. This will initiate your search—see how far you can go!

Learning to Use BLAST

Starting link: http://www.ncbi.nlm.nih.gov

This tutorial is a more detailed look at the BLAST (Basic Local Alignment Search Tool) program at NCBI, the most important single software tool for searching sequence databases. BLAST can be used to search databases using nucleic acid or protein query sequences.

We will walk through two protein examples in this tutorial. The first is the protein insulin, familiar to all of you for its role in the regulation of blood sugar and in diabetes, as discussed in Chapter XX. The second is dystrophin, which we examined in the Entrez tutorial, a large and complicated protein.

To get to the BLAST homepage from NCBI, click on 'BLAST' at the top menu. This page is the starting point for several BLAST programs available from the BLAST homepage. Since insulin is a protein sequence, we will search using BLASTP. Click on **'Protein-protein BLAST [blastp]'** under the Protein BLAST heading.

The BLASTp page that appears contains a window for pasting in the query sequence as well as several optional parameters that can be set.

For our first database search we are going to use the insulin protein from the Zebra fish (*D. rerio*). The sequence of the insulin precursor protein in FASTA format is written as:

>gi|12053668|emb|CAC20109.1|insulin[Danio rerio]

**MAVWIQAGALLVLLVVSSVSTNPGTPQHLCGSHLVDALYLVCGPTGFFYNPKRDVEPLLG
FLPPKSAQETEVADFAFKDHAELIRKRGIVEQCCHKPCSIFELQNYCN**

You should copy and paste the sequence above into the **Search** text box. Alternatively, you could do this search by typing the accession number into the window. Blast will recognize it and retrieve the sequence from the database.

The FASTA format has an identification line followed by the primary amino acid sequence in single letter code. The mature insulin molecule is a processed version of this primary sequence. It is derived from the initial translation product shown here by limited proteolysis.

The remainder of the search form allows you to set various options.

Set subsequence allows you to search with a particular portion of the sequence. Leave it blank so that the entire sequence will be used in the search.

313

Choose database has a drop-down menu that allows you to choose which part of the database you want to search – the entire database (nr for non-redundant), or sections of it such as *Drosophila* proteins only. Leave it on 'nr,' the complete protein database.

Do CD-Search allows a comparison of the query sequence to a database of conserved domain patterns. This is a powerful tool for finding functional domains in genes. Leave it toggled on.

Last, click the **BLAST!** button to start the search.

Pushing the BLAST! button sends the file to the NCBI computer for the search . The page that appears next tells you that the sequence has been submitted and gives you a Search ID number.

Now, click on the image to see conserved domains.

The page that appears shows the result of the **Conserved Domains** search. Conserved domains are the functional modules of proteins. They might include a pattern of amino acids typical of a particular catalytic site, or perhaps the binding site for a regulator of a protein. Click on **IIGF** to see the full list of conserved domains. These will appear in a pop-up window. Our search identifies a domain pattern: insulin and insulin-like growth factor. The colored bars can be activated by rolling your mouse over them, so that the identification of the pattern shows up in the window. Roll over a bar. You will see the text highlighted to give the identification number of the pattern match, the fact that it is for insulin, and a statistical measure of the significance of the match (E). The E value is the expectation that the match would have been found in the database by chance alone (lower is better). In this case, E is very small indicating that our query sequence is likely insulin (but we already knew that!).

You will also see a pop-up describing the domain.

Now, close the CDD window and click the **FORMAT!** button to tell the computer at NCBI to send the results of the actual search.

The output of such a search is many pages long and is made up of several components.

The first major section is a graphical display of the strongest matches to the query sequence. Notice that some hits are to the full-length insulin precursor and some are to the shorter processed form. They are color-coded according to the alignment score. If you roll your **mouse over** the various lines, the identification information, alignment score (S), and E value appear in the text box above the chart.

The second section of the results is a detailed list of hits ordered by their alignment scores. They correspond to the ones displayed graphically. Note that each line gives the identification information for the protein followed by the alignment score and the E value. The entries are ranked from the lowest to the highest E value, which can be interpreted as from most similar to more distant. The top line, not surprisingly, is the record for Zebra fish insulin itself. If you click on the gene identifier link, it will call up the sequence from Entrez. When you click on the Score link, it will show you the particular alignment from lower down in the output file. You can return to this page by using the back button on your browser.

Further down the list, the output gives the actual gene alignments for the various proteins in the list we just looked at. The first of these represents the top alignment. It is a perfect match to the subject, the Zebra fish insulin record. Note however, that early in the sequence is a masked region (shown by gray). This is a low complexity, or repetitive, region that is masked out in the query and ignored in the database search. Such regions may interfere with the alignment. You can see the actual masked sequence, since it is the same protein, in the subject line (LLVLLVVSSVS); it is a mostly hydrophobic, repetitive sequence.

Now let's look at two examples with a less than perfect match.

Scroll down the list of sequences producing significant alignments until you find the *Xenopus* sequence (gi|124513|sp|P12706|INS1_XENLA), and click on its score, '**95.**'

The alignment with *Xenopus* insulin (South African clawed toad) is a less than perfect match. Many regions are still strongly conserved however (notice the low E value). Empty spaces indicate mismatches and a + sign indicates similarity between the two different amino acids compared. For instance, at the beginning of the sequence, a hydrophobic leucine is scored as similar to a hydrophobic valine. Notice also that a gap had to be inserted in the *Xenopus* sequence to give a good alignment with the query. This would be the site of an insertion or deletion event during evolution.

The last example that we will look at is the alignment with human insulin. **Scroll** back up to the list of sequences producing significant alignments, and look for gi|4557671|ref|NP_000198.1|, and click on its score '**82.**'

Notice that there are many records for human insulin in GenBank. The alignment itself is still very strong (note the E value). But, compared to the *Xenopus* alignment, there are several more gaps placed in both the query and subject, not surprising for more distantly-related sequences. Note that the full length of the protein is not shown. BLAST is a local alignment tool and only displays the most strongly matching regions of the overall comparison.

We are finished looking at this sequence, and we're going to search for another sequence. So, **close** the window that the Zebra fish results appeared in, and you should just have the tutorial window open.

For our second BLAST search, we will use the dystrophin protein that we examined in the Entrez tutorial. Dystrophin is a very large protein (several thousand amino acids) with multiple conserved domains. It is also very highly conserved within the vertebrate group.

To run the search let's go back to the BLASTP page by clicking on the **'Protein'** link at the top of this part of the page.

Now, we could call up a dystrophin accession number (P11532), choose the FASTA output and cut-and-paste the sequence into the search window. It is much simpler, however, to simply type in the accession number since in this case we know it and the NCBI BLAST program will accept it.

Try entering **'P11532'** in the Search box and pressing **'BLAST!'**

After the results appear, you should click on the **'conserved domains.'**

This output comes from the Conserved Domains database. This is a database of evolutionarily conserved patterns of amino acids. These small patterns identify the amino acid residues that are almost always conserved through evolution in particular functional domains. There are two types of conserved domains in dystrophin as shown in the graphical representation. The domain at the far left is the calponin homology domain. Calponin is an actin binding protein in the cell. Note the E value in the window by rolling over the blue blocks that say 'CH.' Each database uses a somewhat different pattern, hence the different E values. The remainder of the domains marked are spectrin repeat domains.

Spectrin is an important protein of the cell's cytoskeleton. Dystrophin has many domains showing similarity to a motif in Spectrin. These are smaller and/or have weaker similarity than the calponin domain. Compare the E values yourself. The various domains are allowing you to see the modular nature of proteins.

Now, let's close the pop-up and go **back** to the original results page and click **'Format!'**

Once we receive the search result, it is obvious that there are many similar proteins in the database. Moving the **mouse over** the graphic will display the information line for dystrophin for a variety of vertebrates, a large number of different human dystrophin alleles, and sometimes gene fragments or other proteins aligning with specific domains. Now let's look at the subject hit list below the chart.

Note the E values for many of these subject hits on the right. The expectation is not exactly 0, of course, but rather is very close to a 0 chance of it being by chance alone, and has been rounded off. Note that the first real score is a rather small number, so you can imagine how small the other expectations must be! If we click on one of the scores, we can see the alignment. Let's try the mouse gene by clicking the score of **'1500'** for 'gi|192972|gb|AAA37530.1| (M18025) dystrophin [Mus musculus].'

The mouse sequence is obviously very similar to that of the human but there are lots of positions where they have diverged. Given longer periods of evolutionary separation, we would expect to see more divergence.

We are finished looking at this sequence, and now we're going to try a more targeted search for another sequence. So, **close** the window that the dystrophin results appeared in, and you should just have the tutorial window open.

Let's now ask whether *Drosophila* has a dystrophin sequence.

One way of focusing the search is to specify a smaller database. In this case we can search the *Drosophila* genome alone.

Go back to the BLASTP home page by clicking on **'Protein'** at the top of page. Paste the dystrophin accession number, **P11532**, into the Search text box. Then, lower down, open organism drop-down menu, select **'Drosophila genome.'** Now click **'BLAST!'**

When the results are in, click the **'Format!'** button to get the hit report.

Here the result is not as complicated as for the entire GenBank database. Clearly there are some very good scores, as well as various shorter alignments. Perhaps some of the latter represent calponin or spectrin repeat domains. Let's have a look at the top score by clicking on the **first number** underneath the 'Score (bits)' column.

The *Drosophila* dystrophin gene product is obviously quite strongly diverged from the human version. Here our overall score is very good but the sequence comparison shows only about 40% identities. Nevertheless we would expect, if we examined the conserved regions carefully, that they would represent the important functional domains of the protein. At this level of analysis, we are perhaps not so different from the fruit fly.

You have been investigating the publication record of your genetics professors and searching for information on a gene of interest to them. Go to the NCBI homepage (http://www.ncbi.nlm.nih.gov/), find the BLASTP program and do a BLAST search on one of the genes that you found.

Using BLAST to Compare Nucleic Acid Sequences

Starting link: http://www.ncbi.nlm.nih.gov

In the previous tutorial, "Learning to use BLAST," you learned how to use BLAST to search the genomic databases and to compare protein sequences. Proteins have a high complexity and information content since there are 20 different possible amino acids at each position in the sequence. As we have seen, there are also groups of amino acids with related properties that are scored as being similar. Nucleic acids, on the other hand, have only four possible choices at each position (AGCT[U]). For the most part, we only look to match identical residues at a particular position in a DNA or RNA molecule. In this tutorial we will start using BLASTN to compare the sequences of different transfer RNA molecules.

From the NCBI home page, choose **'BLAST'** from the top menu, and then choose **'nucleotide-nucleotide BLAST [blastn]'** from the list underneath the heading "Nucleotide BLAST."

Transfer RNA molecules have a complex tertiary structure that is essential for their function. It is dependent on intramolecular complementary base pairing and can be seen in the diagram below of the crystal structure of a phenylalanyl tRNA from yeast. The requirement that tRNAs maintain this structure in order to interact with the ribosome and with aminoacyl tRNA synthetases results in strong selection in favor of retaining the primary sequences through evolution.

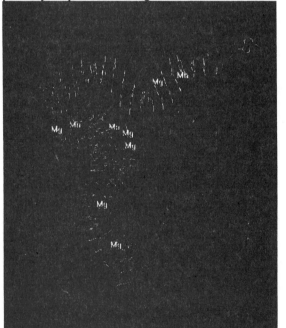

Crystal structure of phenylalanyl tRNA from yeast (MMDB 14286).

The following figure shows the cloverleaf stem-loop secondary structure of a tRNA molecule. This conserved structure is common to all tRNA molecules. In the diagram, we can more clearly see the complementary base pairing in the stems. Ribosomal RNAs that serve a structural role in the ribosome also show strong conservation of some regions. As we will see later, however, the constraints on the structure of messenger RNA sequences are less and there is correspondingly less conservation of sequence.

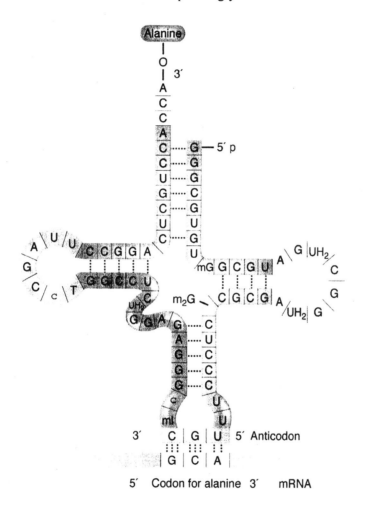

Figure 3-21 from *Modern Genetic Analysis, Second Edition* by Anthony J. F. Griffiths et al., ©2002 W. H. Freeman and Company.

BLASTN is very similar to BLASTP but obviously will only accept nucleic acids as input sequence. Cut and paste the phenylalanyl tRNA from *Drosophila melanogaster* into the **'Search'** text box. In 'Choose Database' choose 'Others (nr).'

>gi|174308|gb|K00349.1|DROTRF2 D.melanogaster phe-trna-2

GCCGAAATAGCTCAGTTGGGAGAGCGTTAGACTGAAGATCTAAAGGTCCCCG
GTTCAATCCCGGGTTTCGGCACCA

Just as for BLASTP, the program is run by pressing the **'BLAST!'** button and then the **'Format!'** button when it appears.

A search against the entire database at GenBank yields many hits, since tRNA molecules are highly conserved. The output list from this search shows many strong hits with low E values. Not surprisingly, many are from *Drosophila*—but also note other species, including human, in the list.

Look at the alignment for one of the human subject sequences by clicking on the Score **'137.'** You should see that it is almost identical to the query sequence.

Now let's look at the source for many of these sequences by clicking on the GI or accession number of any of the sequences. You will find that they may be in very large genome sequence files containing multiple genes; the tRNA subject sequence is somewhere internally in the DNA fragment.

Let's look at more distantly related hits, which would be much further down the list on the previous page. To illustrate a hit with low identity, first you should **close** the window that opened with the first search results in it. Then, click **'Nucleotide'** on the nucleotide-nucleotide BLAST page, so that you can start a new search.

Now, **search** again using our *Drosophila* tRNA, choose NR as above but then choose the **E. coli database** option in the Options where it says 'All Organisms.'

>gi|174308|gb|K00349.1|DROTRF2 D.melanogaster phe-trna-2

GCCGAAATAGCTCAGTTGGGAGAGCGTTAGACTGAAGATCTAAAGGTCCCCG
GTTCAATCCCGGGTTTCGGCACCA

Then, press **'BLAST!'** and then **'Format!'**

The output of our search is similar to the previous one—but note the E values. At this evolutionary distance (*Drosophila* to *E. coli*), there are only a few tRNAs retaining substantial similarity to the query sequence. The graphic display shows that most of these hits are towards the 5' end of our query sequence.

Click on the **'Score'** for the tRNA-Phe (44) gene. Note that it is for about 30 nucleotides of the query. Scroll down and find one where the alignment is to a tRNA embedded in a large genomic sequence. Remember that BLAST is a local alignment tool that finds high-scoring regions within a comparison. It does not usually produce an alignment from one end of the gene to the other.

The first long genomic sequences are in fact the entire *E. coli* genome! Choose *E. coli* K12 and click the score. At position 262095 within the large sequence fragment is a tRNA sequence for an *E. coli* thr-tRNA.

Click on the **GI** or **accession number** to access the Entrez data for this large genome sequence in *E. coli*. Toggle 'Hide' 'Lesser Features,' to OFF. Open 'Features' button and toggle 'tRNA' ON.

Scroll down the data until you see the link for the gene at position 262095. This is the *E.coli* tRNA-Phe gene.

Go back and click on the score (44) for the tRNA-Phe gene.

BLAST does not attempt to do a global alignment from one end of the sequence to the other unless there is an uninterrupted high score throughout the alignment. Our hit reported a match of 31/34 for the 5' end of the molecule. If we align the complete sequence, we see that there are blocks of identity throughout the molecule with a match of 51/76 but only the block from 8-41 gives a high enough score to be reported by the BLAST program (see diagram below).

Remember the E value for this hit (0.0002). This is the probability of finding a match of 31/34 in the database (*E. coli* genome) by chance alone. It is very high because the complexity of nucleic acids is low, having only the four possible base choices at each position. An alignment of two proteins with 31/34 identities would have a much lower E value because there are 20 potential choices at each position and the chance of finding the corresponding matches would be low.

By examining such alignments over their full length as shown below, it is obvious that the subject has significant similarity over its full length.

5'
```
gccgaaatagctcagttgggagagcgttagactgaagatctaaaggtccccggttcaat
gcggatatagctcagttggtagagcagcgcattcgtaatgcgaaggtcgtaggttcgac
                        3'
cccgggtttcggcacca
tcctattatcggcacca
```

Now that we have had a look at highly conserved tRNA sequences, let's try mRNA. Messenger RNA sequences for the same conserved protein drift substantially through evolutionary time. In addition to the drift of the amino acid sequence over time, the degeneracy of the genetic code may have caused the same amino acid to be encoded in multiple ways. Moreover, since the tertiary structural constraints are not as large on mRNA as they are on tRNA or ribosomal RNA, the sequences can diverge quite rapidly.

The p34 cyclin-dependent protein kinase is essential for mitosis and is strongly conserved at the amino acid level throughout the eukaryotic world.

Close the window with the *E. coli* results in it, and then click **'Nucleotide'** from the top of the previous search page.

Now enter the accession number **'M12912'** (for the sequence for the p34 gene from *S. pombe*, the fission yeast) into the Search box, and click **'BLAST!'** and then **'Format!'**

When the results are in, we get a few strong subject hits. The graphic and hit list show strong alignments to several GenBank records for the fission yeast p34 gene itself. Not surprising, they are identical to each other. After that, however, the hits are only for very small regions, scattered throughout the gene and with significant scores in only a few other organisms, mostly other fungi. They do however show clustering in some particular regions presumably associated with conserved domains essential for function of protein kinases.

For comparison, if we run a BLASTP search using the protein product of the fission yeast p34 gene (accession P04551), we would get a very different result.

Close the window with the *S. pombe* nucleotide results in it, and then click **'Protein'** from the top of the previous search page.

Now enter the accession number **'P04551'** into the Search box, and click **'BLAST!'** and then **'Format!'**

Here the hit list is extensive with very low E values and clearly we have found a large number of homologues of this gene.

If we look at the third match (score '412'), we see a highly similar protein in *Homo sapiens*. Here you can see that the protein in the human genome is 66% identical and 80% positive (identities and similarities) to that of the yeast. Furthermore, the similarity extends over the full length of the protein.

There is a lesson in this. If you are searching for matches to a protein encoding sequence, then always search with the translated product.

Now that you have experience doing nucleotide and protein searches, do a search yourself using a gene that is mentioned somewhere in your textbook. Find the accession numbers for the DNA sequence and for the protein sequence in Entrez (make sure both are for the same species and that neither is for a multiple sequence file). Run BLASTN and then run BLASTP and analyze the result.

Learning to Use PubMed

Starting link: http://www.ncbi.nlm.nih.gov

It is the responsibility of scientists to communicate their findings with their peers and also with the world at large. The major means of doing so is the publication of scientific papers and based upon these, textbooks. One of the most difficult challenges for the modern researcher is to keep up with the current research literature.

Researchers must relate appropriate publications to the problem at hand and use the findings of others to help direct their own progress. It is not an easy task. Even if someone had all day every day to sit and read, it would be a formidable accomplishment to say that they had read all relevant material within their own field, let alone that of others. The scientific literature is growing in size so rapidly that computer searches of various types are indispensable. A second aspect of this is that access through the computer increasingly allows direct access to electronic publications.

You have probably already become familiar with searching the databases at your institution's library. In this tutorial we will take a closer look at the research literature database at NCBI. Within the Entrez program at NCBI is the PubMed database, which is produced in collaboration with the National Library of Medicine MEDLINE database. Most importantly, all of the publications at PubMed are cross-indexed to gene and protein sequences and all of the other databases at NCBI. They are also cross-referenced to textbooks in the form of key word searches. Notable in the list of accessible textbooks is Griffiths *et al., Modern Genetic Analysis* (MGA).

Click **PubMed** on the top menu.

We have already looked briefly at this page in previous tutorials. In our introductory tutorial we did a simple author name search for L. H. Hartwell.

Make sure your system will accept cookies from http://ncbi.nlm.nih.gov. Let's try a number of other searches to demonstrate the potential of this system. Let's start with the words in the title of Chapter 6 of MGA: 'Genetic Recombination in Eukaryotes.' First type **recombination** in the search window and press **Go**. The result is a daunting 170,000 plus research articles that mention the word in title, abstract, or key words. It is a popular and important area, but this is far too general a search term to be useful.

How about **genetic recombination**? Try it. We still get more than 150,000 returns. We are retrieving articles actually studying the process of recombination but also many that are of more peripheral interest to us, for instance, simply using recombination as a method to insert genes into chromosomes.

Try the whole title **genetic recombination in eukaryotes**. Now the output is cut to less than one thousand. That number would certainly be far fewer publications than are actually concerned with recombination in eukaryotes. The problem is probably with the term eukaryotes. The key words of publications might include phylogenetic group, but not be as broad as all eukaryotes.

Try **genetic recombination in mammals**. We get more than 110,000 entries. Many of these clearly do not have recombination as the primary subject of the paper.

Let's have a look at how our terms might be related in the formal subject headings within the National Library of Medicine. These Medical Subject Headings (MeSH) are used to classify publications. Click on **MeSH Database** under PubMed Services on the left panel.

At the new screen, type **recombination** in the search window and press **Go**. Next, click on the first item 'Recombination, Genetic.'

Note the definition of the term and scroll down to display various subheadings. This is more like the list of topics relevant to recombination that you would learn in a genetics course. Having a look at these terms sometimes helps narrow your search field by presenting you with some alternative (and narrower) terms. If we searched using the MeSH term *Recombination, Genetic,* we would get more than 100,000 returns. When we look at the subheadings covered we can see why this might be so.

For example, **transfection** is included. Click on it. The definition shows that it includes all papers that use transfection of DNA or plasmids into cells for experimental purposes. Note that it would be included in the techniques tree as well as the recombination tree shown. Many of the papers would be using it solely as a technique and the publication would not be about the process of recombination itself.

Click **Recombination, Genetic**, in the lower tree to go back to our original search.

Now let's try clicking **Crossing Over (Genetics).** This is narrower in scope and focuses on the chromosomal recombination process. Press the **'Clear'** button under the upper search window. Then check the box next to **Crossing Over** then choose **Send to: Search Box with AND.** Press **Search PubMed**. This yields a few thousand entries and a glance at the titles shows us that many are directly concerned with chromosomal recombination.

We could also focus our search by excluding some of the subheadings in the Recombination tree. Open the **MeSH Database** again (from the left-hand menu) and search **Genetic Recombination**; toggle the box and **Send to: Search Box AND.** Then search **Gene Transfer, Horizontal**, toggle it on, and send the item to **Search Box with NOT.**

Next press **Search PubMed**. This returns an extensive list of papers dealing with various aspects of chromosomal recombination but **NOT** gene transfer.

Now click the **Preview/Index** link in the bar under the search text box. This feature shows you a list of the searches you have run and the number of results for each. We could add more terms at this point. Toggle the drop-down menu at the bottom of the page to **First Author** and type **Hartwell LH** in the text box. Press **'AND,'** then **'Preview.'** A new result appears in your list of searches with only a few entries. If you click on this link for the number of results, you will see a subset of Hartwell papers dealing with recombination processes. Notice that there is more than one Hartwell.

There is a lesson here. It is relatively easy to define a narrowly focused search. For instance, we could search **recombination AND Hartwell LH** in PubMed and get a dozen or so journal references. It is often difficult however, to define a broad topic that does not include a substantial amount of irrelevant material (at least to you). If the search is narrowed to keep the amount of irrelevant material low, then it will almost invariably exclude material that you would want included in your list.

This means that you must use a somewhat broader search and then make a judicious choice of material to keep. Let's do a new PubMed search for **chromosomal recombination AND yeast**. If you wish to see the formal MeSH search terms for your search, press Details at the top. (To return to your search results, click on the number under the Result line.)

Our hit list now is down to about a thousand articles, with the titles showing. To assess the relevance of each article, a simple click on each of the authors will give us the Abstract of the paper.

Move down the list and choose a title that you think might be relevant and click on its **authors** to display the Abstract. Go back and forth to find several that meet the criteria of being about chromosomal recombination in the yeast system. For each of these **check** the little box to the left of the reference. When you have found a set, Click **Send to: Clipboard** on the top menus. This will transfer these references to the Clipboard.

When you have finished, click **Clipboard**, on the gray menu bar. This will display only your selections and allow you to collect them by printing or saving to your computer. You could save it as a web page and retain all of the links to the abstracts. Alternatively, you could click Text to convert it to a text file that you can save to your disk and open with programs such as Notepad.

Now let's examine the links to textbooks. **Open** one of the abstracts on your list, and then click on the blue **Books** link under **Links** at the upper right. This searches the keyword lists of the textbooks against the abstract. All words or phrases found will light up as a link. **Open** one. You are presented with a list of textbooks where the topic is covered.

Click on the link to the number of **items** in a specific book to go directly to the list of locations where your keyword is discussed.

Alternatively, you could open the books themselves by clicking on the **book cover**. (You might have to click the very small **Books** link on the top black bar.) Try opening the first edition of ***Modern Genetic Analysis*** to see the relevant table of contents. From here, you can search for a term in the search box to see the paragraphs that will help you define terms and understand the concepts. This will allow you to read many scientific abstracts in areas where you have little experience as yet.

From the book's table of contents you can also **open** the chapters directly. You will see various terms underlined in the text. **Click** on one of these. A glossary appears to provide you with a precise definition.

Initiate your own search by investigating Holliday structures in recombination. You might narrow the search by including a term such as **'sequence'** or perhaps **'genes.'** Use the resources of Entrez to find the sequence of at least one of the proteins involved in resolving the complex.

OMIM and Huntington Disease

Starting link: http://www.ncbi.nlm.nih.gov

We looked briefly at the Online Mendelian Inheritance in Man (OMIM) resource in the introductory tutorial. Now we will explore it more fully. Note that if you are searching for information concerning a particular gene or disease, an OMIM search can be run directly from the initial drop-down menu on the NCBI home page. This is done by choosing OMIM and entering a search term.

However, in this tutorial we want to look at some of the structure of the OMIM database and this information is available from the OMIM home page. Click on the **OMIM** link on the search drop down menu to get to the OMIM home page.

Now let's start with the OMIM statistics page that we saw earlier. Choose **'Statistics'** from under the heading 'OMIM Facts,' on the list on the left.

The statistics page shows the number of loci included in the OMIM database, as well as a breakdown of their distribution by inheritance pattern. Remember that our most recent estimate of the number of human genes is about 30,000. Now, notice that barely perhaps half of all human loci are currently included in this database. There is still a long way to go before they are all categorized.

The various links in the table will call up a list of the genes fulfilling a particular mode of inheritance. Click the first **number** in the X-linked column.

The link takes us to a list of genes in the format of an Entrez search result. The individual records, however, are to annotations within the OMIM database. The Entrez links to sequence and related genes are over on the right side of the page. Let's look at a typical OMIM record by clicking on the first record **'*314998.'**

This record is a description of the discovery of the gene. It includes the literature references and information about linkage and location. This is all within a brief description written by a curator. This is an actively annotated database that changes with time. As research progresses, records of this type must be updated to keep them current.

Now, click on the green **OMIM logo** at the top of the page to get back to the OMIM homepage.

From here, click on **'Update Log'** from under the heading 'OMIM Facts,' on the list on the left to call up a table of the frequency of updates and new additions of records to the database. Note that approximately 80 new records are added each month. Many times that number of existing records are updated to reflect our increasing knowledge of particular genes.

Now, click on the green **OMIM logo** at the top of the page to get back to the OMIM homepage.

Now let's look at a particular case, Huntington disease. Type **'Huntington'** into the OMIM search window, and press **'Go'** to see the display for the OMIM record for this gene. The complete record for such an important and well-studied locus is tens of pages long. We will examine a few of the features.

First, click on the **'Protein'** link (under Link) on the right of the first entry (accession number *143100) to take us to the Entrez protein record for Huntington.

Next, find the entry for accession number NP_002102 either by scrolling through the results, or by doing another search in this database. Then, click on the accession number **'NP_002102'** to see the record.

Scroll down through the human huntingtin protein record until you reach the Comment section. Notice that here is a description of the protein and its expression, the phenotype of mutants and even the nature of the common mutation in Huntington disease. The disease is associated with the expansion of a trinucleotide repeat sequence leading to an enlarged polyglutamine stretch in the protein. Notice also in the text how the NCBI staff indicate precisely the history of this record relative to the initial submissions to GenBank.

From here, scroll back to the top of the page and click on the **'OMIM'** link in the top tool bar.

Next, type **'*143100'** and click **Go** to reach the OMIM entry for this disease and click on the link.

The OMIM record for Huntington disease is a very rich source of information. The research literature is summarized under various headings including Clinical and Biochemical Features, Inheritance Pattern, Population Genetics of the disease, and lots more. Clicking on the various headings on the left will take you to the subsections of the file. This is an immense resource. Entire term papers could be written based on this page alone! Clicking on any of the light bulb icons gives you the relevant literature sources.

Now let's have a look at the gene map. Click on the **'Gene Map'** button on the left.

The map button takes us to a portion of the gene list for chromosome 4 and displays the 4p16.3 region, which includes the Huntington locus, and is listed here by the name of the gene product, huntingtin. Click on **'4p16.3.'**

Now we have the graphical display of chromosome 4 with the Huntington disease locus highlighted. We can see from the graphic of the entire chromosome on the very left that the locus is very near the telomere on the *p*-arm of the chromosome. (Remember that the two arms of a human chromosome are named *p* and *q*.)

By zooming in on the locus (use the zoom in feature just above the chromosome graphic on the very left), we can display the genes that are located close to the Huntington locus. The list on the right is compiled from sequence data. Under the Morbid column are various loci associated with disease phenotypes.

Now, click on **'HD'** under the "symb" column to go to the LocusLink record for the Huntington disease gene.

LocusLink is another annotated database and a very rich resource. Scroll through it. As with all of the other resources we are looking at, it is richly decorated with links to databases and information resources of all types. In the colored boxes are links to PubMed, OMIM, and other databases we have touched upon. However, some are new. Let's click on **'HGMD,'** the human genetic diseases database.

The HGMD database is an outside resource specializing in the nature of characterized mutations for various loci. Note that only the trinucleotide repeat is listed as present for the HD gene.

Now, let's go back to the Huntington disease LocusLink page at **http://www.ncbi.nlm.nih.gov/LocusLink/LocRpt.cgi?l=3064.**

Now, let's go back to the gene map page by scrolling down a bit and clicking on the red link **'See HD in Map Viewer.'**

Now we are back at the map view of the Huntington region. We can go to greater and greater depth from this map, which is largely a linkage map, to the actual nucleotide level. There we can examine the structure of the gene itself. Click on 'HD' in the Genes column and the **'sv'** in the pop-up menu.

'SV' stands for sequence viewer. In this case, it links to a graphical display of the annotated nucleotide sequence for the huntingtin region on chromosome 4. If you clicked on it for other genes in the display, you would activate different regions of the chromosome for viewing. The display shows about a megabase of DNA with several open reading frames indicated graphically. An expanded view of a particular region can be seen by clicking on a particular region of the chromosome and zooming in or out using the tool on the right.

Expand the start of the huntington coding region, indicated by HD at the left, and the three parallel lines under the sequence. The DNA sequence itself is at the top, mRNA below.

Scroll further down along in the sequence and you will see the transcription start site for the huntingtin mRNA. A bit further along, you can see the start of the coding region for the huntingtin protein with the amino acid sequence indicated below. Note the QQQQQ polyglutamine region— this is the region that expands and interferes with function, and is the mutational event that causes Huntington disease. The forward and backward blue arrows at the top of the sequence allow you to walk in both directions along the DNA strand at will. The QQQ region is near nucleotide 1582271.

We have explored OMIM to a sufficient depth for the moment. It is clear that there is much more here. As with all of the other sections of these tutorials you must go to NCBI and examine the links yourself.

Choose a well-defined human genetic disease based on one discussed in your textbook or class. Using OMIM, identify its chromosomal location and determine the nature of the gene product. What is the biochemical lesion involved? Use sequence viewer to find the start and stop sites for the gene. Also find the 5'-end of the first intron in the gene.

Finding Conserved Domains

Starting link: http://www.ncbi.nlm.nih.gov

In the Protein BLAST tutorial we saw how the BLAST site at NCBI first compares the query sequence to a database of conserved patterns or motifs, the Conserved Domains Database. The patterns in the database are designed to detect structural motifs in proteins. By focusing on these conserved domains, such searches help us assign functions to unknown genes and see the domain architecture of complex proteins. We have already seen output from such a search for the dystrophin protein. Let's start there to investigate this feature more thoroughly.

Instead of starting with the BLAST page (where a Conserved Domains search is run as part of the BLAST search), let's go directly to the Conserved Domain Search page.

Click on **Structure** on the top black menu bar, then on **CDD** (for Conserved Domains Database) on the left-hand side bar. (You will have to scroll down to see it.)

Our basic evolutionary model is that sequences will drift further and further apart over time unless selection is at work. When we do a BLAST search, we compare proteins looking for regions of sequence conservation. When we find such evolutionarily conserved regions, we assume that they represent amino acid sequences that are responsible for the compact functional structural features of the proteins.

The order of the various conserved features that might be present in a protein is referred to as its 'domain architecture.' Complex domain architecture is very common among proteins in the human genome and other similarly complex organisms.

The CDD page contains a search entry window as well as information about the program and its underlying databases. Note also that it is closely integrated with the structural databases MMDB, Cn3D, and VAST, which are accessible from the side panel. We will learn to use these in a subsequent tutorial. Keep in mind for the moment that a conserved domain is really a compact structural feature, perhaps a binding site for another protein, or a catalytic pocket.

First, let's look at the nature of a search pattern for a conserved domain. Click on **Pfam**, in the paragraph beneath the search window. This takes us to Washington University in St. Louis, a home of the Pfam (Protein Families) database. Each Pfam search pattern is derived from a multiple alignment of a family of related proteins. The pattern is designed to recognize members of that family when an unknown sequence is searched against the Pfam database.

In the list of options, click on **BROWSE BY** and then family ID.

Pfam is a large database with over 3000 conserved search patterns for all sorts of different conserved domains. The page that first appears shows the top twenty families, a sort of popularity list. Other lists are organized alphabetically. Each line in the top twenty families list gives its name, an accession number within the database, some statistical information, a link to its 3D structure, and a description. This database is an extremely rich source of information about proteins.

Let's try one. Click on the **LRR 1** motif, the Leucine Rich Repeat. An information page should appear telling you about the motif and what its function is thought to be—a protein-protein interaction domain. If you scroll down the page, you can retrieve the protein alignment that gives rise to the conserved pattern. Click on **Get alignment**.

This list is hundreds of entries long. Note that sequences from the various genes (accession numbers on the left) are all a little different, but similar to each other. Also note that the pattern is not a continuous stretch of amino acids but instead is most highly conserved in three distinct clusters. In generating the Pfam search profile, the database combines all of this information to generate a pattern that will recognize all of these entries and others like them.

Click the **back** button on your browser to go back to the information page.

Now try the **View Graphic under Domain Organization** button to the right of the **Get alignment** button. Here we can see the domain architecture for a number of proteins with leucine rich repeats (in green). Notice that different proteins have different numbers of domains and they are located in different locations along the primary sequence. What they all have in common is a conserved primary amino acid sequence in that region, which reflects a compact conserved structural domain with similar 3D structure in each case.

This is a diverse group of proteins. They also have a variety of other functional domains. If you roll your mouse over the domains you can see what they are.

Let's go back to the NCBI CDD page and run a search. Go to
http://www.ncbi.nlm.nih.gov/Structure/cdd/cdd.shtml.

When we run a CD-search, our query sequence is compared to the thousands of stored patterns located in the Pfam database for various protein families (and also the Smart database at EMBL, a related resource).

Let's run a familiar search. As we have already seen, dystrophin has a complex modular structure containing a number of conserved functional domains. Type the accession number for dystrophin (**NP_003997**) into the Search window in the Run CD-Search box in the middle of the page. Leave the Search Database drop-down menu toggled on **CDD**. (If you open this menu you will see that you can restrict the search to just Pfam or just Smart.)

Now press **Submit Query**.

The graphic displays the location of a variety of conserved domains located along the length of the dystrophin protein. It is the result we saw when we ran human dystrophin in BLAST.

Move your **mouse over** the colored domains and watch the display window provide information about the different regions.

Below the graphic is a list of the domains found. It tells you the Pfam or Smart accession number, a verbal description, and the E value for the match. Lastly, there is a detailed alignment for each of the domains found as well as considerable descriptive information about them.

As you can see, with a single search, researchers are able to glean an enormous amount of information regarding the structure and potential function of various regions in the protein.

Let's go back to the CDD search page by clicking the CDD on the black bar on your browser.

By analyzing dystrophin for conserved domains, we have discovered its domain architecture, that is, the order of different domains along its primary sequence. This was visualized as the graphic display with the various features arrayed along it. In principle, functional homologs of our protein would have a similar set of domains arrayed along them. When we did our BLAST searches for similarity, we found proteins with similar primary sequences to our query. In the DART program (Domain Architecture Retrieval Tool) we are able to search for a pattern of conserved domains rather than for the primary sequence itself. Let's try it.

Click on **CDART** at the left side of the screen.

Enter our dystrophin accession number (**NP_003997**) in the search window and press **Search**.

In the result, the graphic of our dystrophin domain architecture is at the top. Below it are proteins with similar domain architectures. If you click on any of the icons on the graphics, it will open a window with the Pfam or Smart pattern information page.

Notice in the graphics aligned below our gene that some of the domains are in similar positions to those in our protein, notably the two calponin domains to the left, and the WW and ZZ domains on the right. The spectrin repeats are more variable, but present. In addition there are other domains that are not present in dystrophin. This program is scoring similar domain architecture patterns and is a very rich source of information.

Note that the overall output of our search is about 20 pages long. **Click** on some of the later pages at the bottom and see some of the more divergent comparisons.

Remember that the Pfam and Smart motifs are identified by running programs such as BLAST to find families of related sequences. The patterns are then distilled from the conserved regions of these gene families. These patterns can now be used to search and identify new sequences and to place them in functional families.

Choose a gene that interests you, either from your text or related to your earlier searches, and run a CD and DART search.

Clusters of Orthologous Groups

Starting link: http://www.ncbi.nlm.nih.gov

All of the tutorials in this set make use of sequence data from the various genome projects. At this point we have the complete sequence for many genomes. We have looked at several ways of comparing genes and proteins and have used these comparisons to infer evolutionary relationships among them. This often helps us to identify the function of a gene because we may have knowledge of the function of a homologue in another species. Typically we have learned the function of such a homologue from a detailed genetic or biochemical analysis of that gene or gene product.

The problem is that most genes have not yet been worked on and are not yet well annotated. We do not know the function of large numbers of them, even in our most investigated model systems. Furthermore, every time we sequence a new genome, we find many genes with no apparent relatives in any system. Clearly, either the databases are not large enough yet or we need additional strategies to determine function.

Lets look at some of the computational comparisons among species with fully sequenced genomes. A number of databases compare proteins across phylogenetic groups. The Clusters of Orthologous Groups (number two on the Hot Spots list on the right of the NCBI home page, click on it) database compares the proteins in fully sequenced genomes. This is a summary table of such comparisons for both prokaryotic and eukaryotic proteomes. Notice the large number of completed genomes and also the number of upcoming ones underway in sequencing centers or completed but not yet included in these tables. Notice that species are given abbreviations or codes for identification in subsequent tables. Click on the *'eukaryotic clusters'* at the upper right. The eukaryotic species are a bit easier to start with since there are only a few genomes included.

The orthologous cluster table groups together different combinations of species. First, scroll down to the bottom of the page. Here is a summary of the total number of proteins examined for each species as well as the number of KOGs (eukaryotic Cluster of Orthologous Groups of proteins). In the last row of the table, ACDHYPE, are the proteins represented in all seven of the species examined. Notice on the right that there are fewer than 1000 proteins included in this set. Click on this link. Scrolling down the list of KOGS shared by all seven species quickly shows you that these highly evolutionarily conserved elements include proteins playing roles in the core genetic and metabolic machinery of the cell. Many of the protein families will be familiar to you from your Genetics or Cell Biology courses. If you click on any individual KOG in the list you will see a list of the various members of the family in each species and a cladogram displaying the evolutionary similarity amongst them. Try a few.

Go back to the Eukaryotic Cluster main page and examine the top of the list. To be included as a KOG (or COG) the protein family must be represented in at least three species. The table starts with these three member families and proceeds down to those proteins included in all species. For instance, ACD—standing for Arabidopsis, Caenorhabditis, and Drosophila—in the top row shows that there are 16 KOGS restricted to these three genomes. If you click on the link it will list the KOGS and the number of paralogues (homologous genes within a species) and orthologues (homologous genes among different species) included.

Go back to the KOG page. A little further down the page, the CDY species group (Caenorhabditis, Drosophila, Saccharomyces) includes only one KOG. Open it and you will see it listed as an uncharacterized protein. We can detect the sequence similarity using tools such as BLAST, but we don't necessarily yet know much about what it does in the cell!

Now go back to the main Cluster of Orthologous Groups page. The unicellular section includes many more genomes. Open the **Unicellular Clusters** link. This is a rich resource including members of the eubacteria and archaea as well as the eukaryotic budding and fission yeasts. Here we can see ancient prokaryotic gene lineages as well as those that were maintained in the evolution of the eukaryotic line.

The database can be examined by species or by COG. For instance if you click on **Mth** (for Methanobacteriales) it will display those COGs found in this group and also list other species sharing them. Try it and then try **Sce** for Saccharomyces (the budding yeast). Saccharomyces is a eukaryote of course and a glance at the types of proteins conserved compared to the prokaryotic groups is mostly the core machinery of the genetic and metabolic system. Many of the protein names should be familiar to you.

Across the top of the page are the letter codes by functional type associated with the number of COGs represented in the species. Click on the first one '**J**' and see that it stands for the translational machinery. These proteins are very ancient in their origins and are shared by most species.

Go back to the Unicellular Clusters home page. Notice the graphic near the top. This is the distribution of numbers of COGs (vertical axis) by the number of species in which it is observed (horizontal axis). Each bar in the histogram can be opened. Click on the first, which displays those COGs found in any three of the species in the database. The list contains many uncharacterized proteins with restricted distributions in the species examined. Now go back and try the bar on the extreme right representing COGs found in all 66 species including the two eukaryotes. Here is the evolutionarily conserved core of our genetic machinery.

Questions

Do you think that the 63 COGs found in all species in the Unicellular Clusters are also in all members of the Eukaryotic Clusters? Which lists might you compare to find out?

Are you able to use this site to identify proteins present in the Archaea but not in the Eubacteria?

Determining Protein Structure

Starting link: http://www.ncbi.nlm.nih.gov

WARNING: Depending on how fast your computer is and how much memory you have available, your computer may be very slow in responding to parts of this tutorial. To speed it up, we suggest closing any other programs you may have open before you start the tutorial.

Genome sequencing has given us an enormous quantity of data regarding the genetic breadth and potential of living organisms. As we have seen, tools such as BLAST make it relatively easy to search the genome databases using the primary sequence of hypothetical proteins. This allows us to ask questions about similar proteins, homologs, and gene families. The primary sequence determines the three-dimensional shape of a nucleic acid or protein. Structural databases such as Entrez Structure at NCBI contain details of protein structure based on X-ray crystallographic and NMR studies. From this database it is possible to retrieve and examine a particular three-dimensional structure in a viewer called Cn3D. In this tutorial, we will examine some structures and learn to use the Cn3D structural viewer.

First, let's download the Cn3D viewer plug-in for your browser. Click on **Structure** at the top blue bar of the NCBI homepage. Then, click on **Cn3D v4.1** on the left-hand side menu.

Cn3D provides a graphical view of the three-dimensional structure for a molecule. Follow the directions to download the program. First click on the appropriate operating system for the computer you are using (**PC, Mac, Unix**) on the blue bar at the top of the page. Read the system requirements and directions carefully. In some cases (for PCs) there is more than one version of the program. Either allow the file to open automatically, or save it to your disk and then click on it to run. The program will install it self and will recognize structure files automatically when we later download them.

Now let's go back to the Structure page by clicking the **Structure** heading at the top of the page.

The structure database can be searched through Entrez. In the search text window, type **1JM6**. This is the identification number for rat pyruvate dehydrogenase in the PDB protein database. Click **Go**. When the result appears, click **1JM6** to see the NCBI Molecular Modeling Database (MMDB) Structure Summary page.

This page provides a description of the file including the MMDB ID number, PubMed link, species, and citation information. Notice a large number of viewing options at the bottom of the page. The default settings are set to automatically launch the Cn3D viewer when the structure file is downloaded. Leave them at the default settings. Make sure that you downloaded the viewer, then click on **View/3D Structure**.

You will see the file download and then two pop-up windows for Cn3D will appear. Move them to the left if need be so that you can still follow this text. Tell it to Open if a pop-up appears.

The upper pop-up window (Cn3D) shows a representation of the 3D structure of the protein. This is drawn from the 3D crystal coordinates for each of the atoms in the molecules. The representation is in structural mode: alpha helices are shown as green cylinders and beta pleated sheets as gold flat arrows. The connecting amino acid chain is in blue.

The display is interactive. The molecule can be rotated in three dimensions. Hold down the mouse button and drag to rotate the molecule. Try it. The calculations are done on your computer and the rotation speed will be affected by how fast your computer is. As it rotates, note that it consists of two mirror image complexes consisting of the A and B chains.

As the structure rotated, did you notice the bound ADP molecule in each complex (partly colored in red and complexed with a Mg ion)? Can you identify the adenine ring structure in the ADP molecule? You may enlarge this window to full screen to make it easier to see, but when finished, reduce it again. Alternatively there is a Zoom In feature under View in the top menu.

The lower pop-up window (Sequence-Viewer) shows the primary sequence of the two amino acid chains. Marked along its length is a representation of the helical and pleated sheet domains. Notice that the primary sequence is color-coded to match the representations in the 3D viewer.

Now go back to the Cn3D viewer window. The representation is of secondary structure. On the menu at the top, open the **Style** drop-down menu and **Rendering Shortcuts** then **Wire** to see a different representation of the structure of the amino acids chain. It is still color-coded for the different structural features. Lastly, try **Spacefill** under Style. (This model requires much more computation for each view). This is a more realistic view of what the surface might really look like, if you were another molecule! Try rotating it.

Close the pop-up windows and let's look at another example. Click **MMDB Structure** at the top of the page to go back to the homepage.

The Structure database is linked to Entrez. Therefore we can enter terms and search in the same manner that we have done in other tutorials. Enter 'GCN4 AND yeast.' This protein is a transcription factor that binds DNA. Click Go.

The list of hits is four pages long and represents all of the entries in PDB for this protein, its fragments and various mutants. As with all of the databases that we have looked at, it takes a certain amount of sorting to get what we want. Go to the bottom of the list. (Select page: 4 in upper right corner then scroll down the page). Choose the last entry, 1YSA, by clicking on it. The MMDB structural summary for 1YSA is shown. Click on the View/3D Structure button.

As before, the two Cn3D pop-up windows should appear.

This view shows us the interaction of chains of GCN4 transcription factor (in green) bound to a promoter site on DNA (in red). We are initially looking down the length of the DNA helix. First, let's switch the view to WireFrame in the Style menu at top. This will let us see the DNA base pairing. Now rotate the image through about 90 degrees. Stop when the red/blue nucleic acid chain is vertical. As it rotates we can plainly see the DNA double helix with its planar base pairs (in blue). We can also see the GCN4 chains contacting the DNA strands on each side. This is the sequence-specific recognition of the DNA molecule by the transcription factor. We can also see that the protein chains interact with each other at their opposite ends.

Now take a look at the Sequence-Viewer. If you use your mouse to highlight a part of the nucleic acid or protein sequence (whichever you have showing in the panel) then it will color the corresponding sequence in the 3D graphic. Try it.

Lastly, place the viewer in Spacefill mode to have a more realistic sense of what the molecules are really like. (Again, depending on the speed of your computer, this may take some time, though not as much as the first protein, as this is a much smaller molecule.)

Close the viewer windows, and go back the Structure homepage by clicking MMDB Structure at the top of the page.

In addition to allowing an investigator to search for the known structure of a particular protein, NCBI also allows an investigator to search the database for structures similar to the one of interest. This allows a crystallographer with a newly acquired sequence to search for structurally related proteins utilizing the VAST program (Vector Alignment Search Tool).

Within VAST, all of the structures in the MMDB database have already been compared to all other structures in the database. Since we don't have a new crystal structure to try, let's explore the structural comparison function using a previously known sequence. Type the PDB identification number **1W98** in the structure search window and click **Go**. This displays a record for a human cyclin dependent protein kinase, one of the proteins that controls the cell cycle. Click on **1W98** in the record. This takes us to the MMDB Structure Summary page for this protein.

First let's click on **View/3D Structure** to see what this protein looks like. **Close** the Sequence-Viewer window, and minimize the Cn3D viewer window for the moment.

The Structure Summary page includes a line labeled Structure Neighbors 3D domains: 1, 2 for each of the chains. These symbols refer to the whole chain of the protein or the subdomains of it. These links take us to the most structurally similar files in the database. Click on **1** in the upper polypeptide, a serine/threonine kinase.

A VAST Structure Neighbors summary page appears. Notice that it has a table of records with varying degrees of amino acid identity. Check the top one (click on the small **square** to the left). This record is for another protein kinase found in humans. Now click on **View/3D/Alignment**. A new Cn3D window will open. This contains a comparison to our original **1W98** file. Both structures are compared.

Enlarge the original Cn3D window containing **1W98** and arrange it along side of the new window.

The new window contains both the 1W98 and the new structures superimposed on each other. You can highlight the two sequences by using your mouse to highlight portions of the sequence in the alignment window DDV at the bottom. You can see that one of the sequences in the 3D model turns yellow. If you look carefully you will see that close to the yellow is a second chain directly paralleling the yellow one. This is the second sequence. If you now use your mouse to **highlight** this region on the second sequence, you will see the corresponding change on the 3D image. The VAST program statistically compares how closely the two models fit to each other and aligns them in three dimensions.

Programs such as VAST are capable of finding very distant homologs, even where we do not see statistically significant sequence identity using alignment programs such as BLAST. Ultimately, it is the 3D structure of a protein that is selected during evolution. It is possible to have similar 3D structures, but quite different amino acid sequences.

Close all of the pop up windows.

342

This is just a brief look at the richness of the 3D information that is available. Such information is critical to researchers wanting to understand how proteins interact or how a drug inhibits a protein.

Go back to the NCBI Structure page and search in Entrez Structure for a structure that interests you. It could be one of the sequences used in earlier exercises or something new. Keep in mind that only a small proportion of all sequences in GenBank have had their structure determined. As a start, you might try looking up the structure of a transfer RNA.

Functional Analysis

Starting link: http://www.ncbi.nlm.nih.gov

In the Clusters of Orthologous Groups tutorial you saw that we can identify families of proteins by sequence similarity but that often times we do not know the biochemical function of the group. Once a researcher identifies a function for a member of a particular family in a species then we can infer (since we have concluded that they are homologous) that the function of the other members of the family is likely to be the same or at least related.

How do we gather functional information? There are not enough researchers in the world to assign one to each unknown protein in the database and tell them to figure it out. We need other strategies. Let's look at some high throughput methods.

A geneticist links function to a particular locus by examining the phenotype of a mutation in that gene. From the change in phenotype (typically loss of function) caused by the mutation we try to deduce what the wild-type function of the gene product might be. Over the years the genetic map becomes more complex as new mutations are added. However, even after almost 100 years of fruit fly genetics, we still do not have a mutation for every gene. Furthermore, many genes when mutated do not generate a detectable phenotype. What else might we try?

With the complete sequence of a genome available we are able to target mutations to particular genes using the techniques of transposon mutagenesis and homologous recombination. This is relatively easy for prokaryotes and yeasts but progressively more difficult as we move to more complex organisms. In yeasts for example, we are able to make a set of yeast strains each of which has had a single gene deleted. We can thus ask about the function of each gene in the entire genome.

Do a PubMed search for two authors: G. Giaever and M. Johnston (enter as giaever g AND johnston m).   This finds two publications with many authors. They are the result of an international consortium that undertook to delete all yeast genes and also to do a phenotypic analysis of each strain. If you click on the 2002 paper you may read the abstract and get an overview of the project and results. Note that as each gene was deleted by homologous recombination it was replaced by a selectable marker and also a unique sequence tag (molecular bar code). The bar code allows each strain to be identified by hybridization of its DNA to a microarray containing all of the bar code sequences. Such deletion strains can be tested individually for changes in phenotype. They can also be screened collectively using the microarray to detect the survivors of a screen. For instance, all 5000 or so strains can be grown in media with high osmotic potential (e.g., high salt concentration), and the survivors detected by isolating DNA and hybridizing it to the bar code microarray.

Such phenotypic information is linked to genes and by extension homologous gene families within the genome databases. Let's look at an example. The Saccharomyces Genome Database (http://www.yeastgenome.org) compiles information on all budding yeast genes. Go to this site. Type **"hog1"** in the search window at the top and click **Submit.** This page summarizes all data concerning the gene and the interactions of its gene product. The Hog1 protein in budding yeast is necessary for osmoregulation. Scroll down the page and click on 'HOG1 Phenotype details and references.' In the table note that there are several phenotypes listed based on data in the Giaever et al., 2002 systematic deletion study. The gene is non-essential but clearly necessary for an appropriate response to a variety of conditions. If you click on any of them, the database will display a list of all other genes in the yeast displaying a similar phenotype.

Scroll further down the page to 'Genetic Interactions.' Here you can find all known genetic interactions between mutations in the HOG1 gene and mutations in other genes in the genome. This type of data allows a researcher to formulate hypotheses about other members of the HOG1 and related biochemical pathways.

Even with a full gene deletion set available we still have many genes in budding yeast with no known phenotype when deleted.

How else might we gather information?

Proteins with related functions often physically interact with each other in the cell. They may be different subunits of an enzyme complex for instance. Defining physical interactions is another powerful way to investigate the role of various genes and to infer the function of uncharacterized ones. Use the Back button on your browser to go back to the HOG1 summary page. Scroll down and click on 'HOG1 Physical Interactions."

This page summarizes physical interactions as shown by co-immunoprecipitation, affinity capture, and two hybrid interactions. Each of these methods detects proteins physically bound to the HOG1 protein and thus we can hypothesize that they may have related functions. Each one can be explored by clicking on it (try one or two and then return to the current page). Each of the interacting genes or gene products has a list of its own interactions actions (which in this case would include HOG1). The interacting network (interactome) builds out rapidly and eventually when we have all the data, will encompass all genes in an organism. Obviously the complexity of this network is difficult to visualize.

Notice that many of the interactions listed in SGD give BioGRID as the database source for the information. **Click** on **BioGRID.** A pop up will appear with a search window. The BioGRID database is built from interaction datasets.

Type HOG1 in the search window. Toggle the organism to Saccharomyces and click **Submit.** The search retrieves all of the data and references that we saw in SGD and also provides a graphic of the interactions with HOG1 as the node. Click on JPG under the graphic to enlarge it. This gives some sense of just how complex the interactions of a single gene/protein are in a cell. HOG1 is an important gene and plays a major role in the cell's economy.

Keep in mind that each of the 63 interacting genes shown would in turn have a similar set of interactions of its own. If you called each of them up, it would be at the center of the node with its interacting proteins surrounding it. Click on **PTP2** (second item) in the list of interacting genes on the HOG1 search result page. Enlarge the graphic by clicking on **JPG.**

Note that there are far fewer interactions for PTP2 but that HOG1 is one of them. It may be that PTP2, a phosphatase, interacts with fewer proteins. Alternatively, we may have incomplete information regarding its role, and more interactions will be found with further work. These databases are dynamic and are constantly being updated.

Go back and try several more of the links to proteins shown to interact with HOG1 (include CRM1 in your list).

If your institution has access to *Nature* then you can see a larger part of an interactome network here (http://www.nature.com/nature/journal/v415/n6868/fig_tab/415180a_F2.html).

Imagine having all of the links for all proteins in the cell in a single searchable diagram. Imagine visualizing all of the interactions for the more than 5000 budding yeast genes at once in a dynamic graphical display. Such resources exist but are not readily displayed here.

Exploring the Cancer Genome Anatomy Project

Starting link: http://www.ncbi.nlm.nih.gov

A number of specialized databases are available through NCBI. These often represent collaborative efforts between outside resources and NCBI. We have already made use of several of these, for instance in the Conserved Domains tutorial and in using OMIM. Many others are listed on the right hand side of the NCBI homepage. The Cancer Genome Anatomy Project is organized by the National Cancer Institute. NCBI provides a number of database and bioinformatic resources to the project. This allows cross-linking among all of the related databases.

Cancer is a very complex disease. In general, it is characterized by inappropriately controlled cell proliferation. The resulting cells interfere with other tissues and they characteristically spread in the body. There are many ways in which a cell's machinery can go awry and still have this overall end result. Cancer, therefore, can be thought of as hundreds of different diseases.

Most cancers arise from spontaneous somatic mutations accumulating over your lifetime. In this sense it is a disease of aging, although it too frequently occurs in younger people. Somatic mutations, of course, are not passed on to the next generation. There are also cancers caused by inheritable germ line mutations and they can be followed through families by pedigree analysis. Characterizing all of the genes and mutations that directly cause or predispose us to cancer is the goal of the Cancer Genome Anatomy Project.

Click on **Cancer Genome Anatomy Project** (NCI_CGAP) in the right-hand menu at the bottom.

This page contains a description of various aspects of the NCBI-CGAP collaboration.

The Cancer Genome Anatomy Project (CGAP) pulls together a large amount of public data on genes altered in cancer cells. This ranges from gene sequence, mutational analysis, cytogenetic localization, chromosomal aberration, biochemistry, and clinical manifestation for cancers. By making all of this information easily accessible, CGAP helps the researcher analyze new information and relate it to what is already known. Since cancer ultimately is a problem of expression of a mutated gene product(s) and consequent inappropriate expression of other genes to give the cancer phenotype, analysis of gene expression has been one of the major efforts of CGAP.

Click on the **Cancer Genome Anatomy Project** link.

This is the homepage for CGAP (http://cgap.nci.nih.gov/). Notice that we can enter the databases from several directions and that they are extensively linked with NCBI at various levels. We could ask about gene or tissue expression patterns, chromosome aberrations, biochemical pathways and various other resources.

Let's try several. First, click on **Genes**.

The GENES page provides access to a number of bioinformatics tools.

Click **Gene Finder** in the top list.

Gene Finder allows us to search the database by gene name, accession number, or tissue where a gene is expressed. The key words and identifiers are from the UniGene classification at NCBI. Let's try an example. In the lower search window, toggle the Tissue Type to **Mammary Gland/Breast** and then type **BRCA1** in the Keyword in Gene Name textbox. BRCA1 (breast cancer associated) is known to be involved in some forms of early onset breast cancer and occurs in families as a germ line mutation. Press **Submit Query**.

Gene Finder returns the BRCA1 gene and a number of records for it. It also returns records for a number of genes whose gene products are known to interact with the BRCA1 protein. The **Gene Info** link associated with BRCA1 takes us to a summary page containing the various GenBank accession numbers as well as links into various databases. Click on it.

The Gene Info page is a compilation of information and links concerning our gene. Scroll down the page and look at the scope. Near the top are a number of Database Links. Click on **UniGene**. The pop-up window that appears gives a summary of information about the gene and about the various GenBank records for it. Note the related proteins in a variety of eukaryotes, even plants. Remember that cancer is a disease that affects the basic cell proliferation machinery and much of this is conserved in all eukaryotes.

Close the pop-up window.

Clicking **OMIM** takes us to a long list of entries on the nature of the gene and its inheritance. We have looked at some of these resources before. It would take many hours to follow all of the links and text.

Remember that these databases could also have been found by entering BRCA1 in Entrez.

Now let's go back to CGAP and see what other types of information it might offer. Close the pop-up. Click on the **CGAP icon** at the top left of the page.

As we unravel the biochemical function of various genes and gene products, we can place them in functional pathways using all of the information available to us. Click on the **Pathways** link.

Click on the **BioCarta Pathways** link.

BioCarta has an extensive list of biochemical pathway charts.

Under A, click on **ATM Signaling Pathway**.

Notice the schematic of the cell cycle across the top of the map. The ATM pathway is important for delaying cell cycle progression to allow time for repair if there has been damage to DNA. This helps to prevent permanent damage or chromosome breakage and the consequent genomic instability. Maps of this type help the researcher place a particular gene product into a broader context. Our BRCA1 gene product is shown just to the left of center. Its precise role, since mutations in BRCA1 lead to a high risk of breast cancer, is the focus of much research around the world. Clicking on any of the icons in the map takes us back to the Gene Info page for that gene. You might have noticed, when we examined it, that the bottom of the BRCA1 Gene Info page contained a link to this map.

Let's go back to the CGAP Genes page and try another link. Click on **Genes** in the green menu near the top of the page.

One of the problems facing all researchers (and students) is the development of a formal language describing the function of genes. This is essential if software is going to be developed that will allow us to move transparently among a variety of different databases. An 'ontology' defines function in a hierarchical fashion, moving from the general to the specific. Click on the **Gene Ontology** (GO) browser.

The pop-up page that appears contains a very general opening page with three descriptors: Biological Process, Cellular Component, and Molecular Function. The + sign in the boxes indicates that there are hidden trees for each.

Click on each of the **+ signs** next to the three terms in turn and open up the next level of the hierarchy. Each line starts with a symbol.

Try clicking on the + sign next to 'nucleic acid binding' in **Molecular Function.** This opens up further categories including whether gene products bind DNA or RNA. Click on **Nucleic Acid Binding.**

Click on the + **sign** next to 'DNA Binding' and further categories open with increasing levels of specificity.

If you click on one of the numbers it will call up a list of the genes in the main browser window. This pop-up window will go to the background and you will have to click on it in your tool bar to bring it to the foreground again.

In some ways working through this hierarchy is the opposite of what we have done in previous tutorials. We have started with a gene and built our information tree from the gene up, until we get clues to function. In the hierarchical structure of the Gene Ontology we narrow the functional fields and then ask "What genes belong to this group?." These terms and the hierarchical structure play a major role in categorizing the genes of organisms as new genomes are sequenced. The terms provide a common language for comparison. Keep in mind that BLAST scores are a major criterion in deciding whether a newly discovered sequence has a similar function (and ontology) to a protein whose function has been previously investigated.

A gene product may belong to more than one category. This is especially true of human genes where very complex domain architecture often correlates with multiple functions.

Let's go back to the CGAP/Gene home page. **Close** the Gene Ontology Browser, and click on **Genes** in the green menu near the top of the page.

We now know of several million different single nucleotide polymorphisms (SNPs) in the human genome. In the case of hereditable disease, such as BRCA1 associated breast cancer, we can track the alleles responsible. Most importantly we can use the information for diagnostic tests to assess the risk to individuals. Click on the **SNP 500 Cancer** link.

Again, a pop-up window will appear. Type **BRCA1** into the gene symbol search window and press **Submit**. A list of SNPs for the BRCA1 gene and some BRCA1 associated proteins appears. Notice in the column at the right that many SNPs do not cause changes in amino acid sequence in the gene product. Click on the accession number of one of the BRCA1 mutations that does cause a change. A graphic appears of the sequence and the change (e.g., Pro871Leu is a change from proline to leucine at amino acid position 871).

Let's **close** the pop-up window and go back to the CGAP main page and examine other resources by clicking the **CGAP icon** at the top right of the page.

Next, let's try clicking the **Chromosomes** link.

Several resources are listed. Click on **Genetic and Physical Maps** at the bottom. A new pop-up window will appear.

Click on **SNPMAP** and **LinkageMap** for **Chromosome 1 linkage** on the menu. This yields graphics of chromosome 1 as well as information on what has been mapped. If you click on various regions, it shows you the physical mapping information for all known genes in the region as well as various ESTs that have been mapped. Such detailed mapping information is instrumental in compiling and interpreting the human genome sequence. It is of tremendous value in detailing the nature of chromosomal rearrangements that are often common in cancer cells.

Now let's **close** this pop-up window and go back to the CGAP homepage by clicking the **CGAP icon** on the top right of the page.

CGAP aids the researcher both by being an archive of information as well as being actively engaged in the compilation of new information and resources. Start your own search through the system. Perhaps begin at the BioCarta map for Cell Cycle: G1/S checkpoint. Checkpoints delay cell cycle progress while DNA repair occurs and many of the proteins in these pathways are associated with cancer. Find the p53 gene and click on it. This gene is mutated at high frequency in human cancers and you will find a wealth of information about its function.

Measuring Phylogenetic Distance

Starting link: http://www.ncbi.nih.gov

Natural selection depends upon heritable variation within a population. The variation is a result of random mutation as well as the random choice of gametes. The net result is phenotypic differences upon which natural selection can work. Differential reproductive rates based on these phenotypic differences result in changes in the genotypes represented within the population over time.

It follows that over long periods of time, random change and selection will lead to increasing DNA sequence divergence among organisms. With certain underlying assumptions, we can use this information to estimate evolutionary distance. We have already looked at this to some extent in the BLAST tutorial and also when considering the nature of Conserved Domains. In this tutorial, we want to explore the relationship in a bit more detail and to examine some of the available resources.

NCBI is the repository for vast amounts of sequence data. It currently has at least one bit of sequence from each of more than 165,000 different organisms. By comparing homologous genes among different organisms we can get a measure of sequence divergence and from that, relative estimates of evolutionary relationship. Different genes, however, may diverge at different rates, so evolutionary relationships inferred from limited data must always be treated with a certain amount of skepticism. Clearly these data must also be combined with insight gained from comparisons of phenotypic characters and the geological record to reach a consensus within the scientific community regarding the relationship of one organism to another.

First let's look at the taxonomy resources at NCBI. NCBI has undertaken to generate a consensus taxonomy linked to all of the sequences in the database. Obviously sequence data plays a major role in defining such relationships. First let's look at the NCBI taxonomy resource. Click **TaxBrowser** on the upper menu.

This page is the gateway to a variety of resources. One of the goals of the Taxonomy section at NCBI is to generate a consensus Taxonomy for all of the organisms represented in the database. Let's have a look at one of our most commonly used organisms. Click on the **Drosophila melanogaster** link in the organism list.

This opens a summary page of taxonomic position as well as a list of the number of records of various types in the database. The lineage is abbreviated in this representation. Click on **Drosophila**, near the end.

There are a lot of fruit fly species in the world! On the upper gray menu, **toggle** the **proteins** to on and press **Go**. This will display the number of sequence records available for each item. Notice that for some, there is only a single entry.

Search down the list and find the melanogaster group (under Sophophora), melanogaster subgroup and eventually *D. melanogaster* itself. Each of these species has a unique genome. Our premise in looking at sequence data is that it would diverge more and more relative to *D. melanogaster* itself as we went further up this taxonomic tree. The quality of our comparison is dictated by the amount of sequence available for each species and the extent of our comparisons. Although we have the entire genome for *D. melanogaster*, that is not the case for the others.

To use sequences in order to help build a phylogenetic tree, we must align the sequences with one another to score their divergence.

In the BLAST tutorial, we saw that more distantly related organisms had more divergent sequences. When we examined the output from a BLAST search, the hits were ordered from most similar to most divergent and ranked by the BLAST score.

BLAST results are a complicated list of accession numbers and formal species names. It is often difficult to understand the relationship of the various organisms to each other simply by looking at the list. Many times, without expanding each entry, we cannot tell what organism we are looking at.

BLAST results, however, are linked to the Taxonomy browser. Let's run a BLAST search and explore it. Click on the **NCBI icon**, then **BLAST** and **Protein BLAST**. We will search using an *Arabidopsis* PCNA sequence. This protein is part of the DNA synthesis machinery and is strongly conserved in all organisms. Enter the accession number **NP_180517** in the window, press **BLAST!** and then press **Format!**

As we have seen before, the BLAST results page is a detailed list of hits, but with limited visible information about the organisms in the list. Clearly we can open each entry and examine them one by one. However, BLAST results are linked to the Taxonomy Browser. Go to the left of the page (above the graphic display of results) and click on **Taxonomy Reports**.

The BLAST result is now formatted in the new window with the lineage of *Arabidopsis*, our query organism at the top, and then each hit on the list arrayed according to phylogenetic position.

Clicking on the **Organism Report** link at the top takes us down the page to list the hits from each organism.

Similarly clicking **Taxonomy Report** gives us the consensus taxonomy with all of the hits listed in it.

This is a very useful tool for orienting yourself to the strange organisms that we deal with.

Close the pop-up windows.

It is possible to use the quantitative measure of sequence divergence to construct phylogenetic trees. The underlying assumption is that the sequence divergence is related to the evolutionary distance.

We will build a small phylogenetic tree based on a sequence comparison of a single protein that is strongly conserved in all organisms. The protein is PCNA, which we used above, and which is part of the DNA replication machinery. We will use the ClustalW program to generate a multiple sequence alignment and then the Phylip program to form our tree. These two programs are not available at NCBI, so we will run them at the European Molecular Biology Laboratory (EMBL) site, one of the archival sequence databases centers in Europe.

We have looked at sequence file formats in earlier tutorials. For this exercise we need to construct an input file for the ClustalW program. We will do this as a multiple FASTA file of PCNA sequences chosen from a number of different organisms. This can be done by collecting several different PCNA sequences on the Clipboard at NCBI, displaying them in FASTA format and saving the result as a text file (.txt) that can be edited in Notepad or a similar text editor.

Let's choose some sequences. Go to the **NCBI** homepage by clicking the icon at the top left of the page. Then, go to **All Databases**, toggle the search to **Protein** and enter the accession numbers CAC_27344 then click **Go**. Toggle box and **Send to Clipboard**. Repeat the search for *Arabidopsis thaliana* (AAC95182), *Drosophila melanogaster* (NP_476905), *Danio rerio* (NP_571479) and *Gallus gallus* (BAB20424), and after each search **add** these sequences to your clipboard. At the end you should have five sequences on your clipboard.

Click **Clipboard** and see your sequences. It should display the summary for each of the five records. Toggle the Display window to **FASTA** and Click **Display**. Now you should see each of the files in FASTA format. When converted from its current HTML format to text format it will be a suitable multiple FASTA input file for a multiple alignment program.

We have to export this back to our own computer as a text file. Click **Text** on the gray toolbar. Now the file consists of identification lines (starting with the > symbol) followed by the sequence in each case. There are no spaces between the entries. This file is now in the proper FASTA format. Under **File, Save As**, name it and save it to your hard drive or disc as a text file (be sure to change the Save as Type drop down menu to Text File).

Open the file with Notepad or a similar text editor program on your computer. Towards the end of the identification line for each sequence is the name of the organism in brackets. We want to keep this information and eliminate all of the accession numbers, etc. For each of the identification lines edit the information behind the > symbol until only the name of the organism remains. This information will be printed out on our phylogenetic tree and shorter is better.

To run our multiple alignment Click **here** to go to the ClustalW program at EMBL at http://www.ebi.ac.uk/clustalw/#.

This is a somewhat complicated web page, but we do not have to concern ourselves with all of it. Most of the windows can be left on their default settings. We do have to fill in several of them, however, and load our sequences.

First, put your **email address** in the upper left-hand window so they will know who you are and give your alignment a **title** in the adjacent box.

In the fourth row, toggle the Tree Type to **None**.

Then push the **Browse** button at the bottom and locate your PCNA text file on your disk. This will upload your file to EMBL (alternatively **Copy** and **Paste** the file from Notepad).

Last, push **Run**.

Click on the Alignment link (.aln). The most similar sequences are on the top and the most divergent on the bottom. Not surprisingly, the chicken and human are closely related and on top and the plant sequence is at the bottom.

Below each column the (*) identifies residues that are identical in all five sequences and the other symbols give degrees of similarity. Close the pop-up.

An alignment of this type is the starting point for doing the phylogenetic tree. To build a tree, a program such as Phylip quantitatively scores the similarity between each pair of sequences based on this alignment. In this case, the tree building program has already been run.

Go back (click on Submit another job) and run the program again. This time toggle Tree Type to **Phylogram**. If you **scroll** further down the results page, you will see a simple diagram displaying this information in graphical form. The length of the lines between organisms and branch points are related to the degree of similarity between the sequences in the alignment. Based on our assumption of constant mutation rate over time, these distances are then proportional to evolutionary distance.

You should **save** this output page to your own disk or print it for your use.

This has been a simple exercise to look at Taxonomy resources and the basics of multiple alignments and their use in tree building. Keep in mind that there are many assumptions underlying such exercises and many alternative ways of doing this type of analysis. Ultimately we satisfy ourselves of the validity of the result when multiple methods give us a similar result. For instance, we might expect to get a similar tree if we used a different protein from the same set of organisms.

Try running an analysis on your own. Choose a protein and make a new multiple FASTA file. You might choose a different protein for the same set of species that we just analyzed. Do you get an identical tree?

Environmental Genomics

Starting link: http://www.ncbi.nlm.nih.gov

Inherent in all of our discussions of genomics is the idea that we can examine the genes/genome of various organisms and from that infer something about the organism's biology. As you have seen in some of the other tutorials we use sequence alignments to identify families of genes or proteins and then make use of databases (such as those at NCBI) linking these genes to function in the broader scientific literature. We now have the complete genome sequence for most of our major model genetic systems as well as for many other Eukaryotes, Eubacteria, and Archaea species. Each year we see the completion of many more as genome sequencing projects are finished. In addition to the several hundred complete genomes, the databases also contain partial sequence for over 165,000 organisms at the present time, and the entire database is growing rapidly. The growth statistics can be seen at http://www.ncbi.nlm.nih.gov/Genbank/genbankstats.html. The database currently contains a staggering 100 billion base pairs (100 Gigabases) of data. These data contribute to our understanding of the phylogenetic relationship among organisms as you saw when using the Taxonomy link.

Large scale sequencing centers with their highly automated systems have allowed us to collect these datasets. They have also allowed researchers to turn their attention to a new question: What is the extent of microbial diversity on the planet? And how does it change with environment? Let's examine how we can monitor such things.

How can we detect different species using sequencing technology? Click on the NCBI icon and go back to the home page. One of the main links at the top of the page is **TaxBrowser.** Click on it. This page displays links to many of our major genetic model systems. As well, in the text section at the top, it gives access to the top level of the taxonomy database through the *tree* link. Open it.

Open the Eukaryota section in the list at the top. If you scroll about one third of the way down the page (or use the Find on this Page function in your browser) you will see a small section for the Metazoa which includes the familiar Vertebrata if you open the links and follow them. The only point to make here is that at the higher levels of organismal diversity in the Taxonomic Tree, the Metazoans are a very small component. Backup to the top of the first page (before searching for the Metazoans). Here we see the *Acanthamoeba* group, some of which are opportunistic human pathogens. Amoebae should be familiar to you from introductory biology. Open the link for *Acanthamoeba castellani*. This is a common species of amoeba and you may have seen it in a laboratory. The page opens to summary tables of the total sequence data available for this species and over on the right a summary of the entries in various parts of Entrez. Notice the more than 30,000 records for nucleotide sequence. If you opened it you would see a long list of sequence records. This organism is clearly substantially sequenced in its entirety and we have a wealth of genomic information available. Go Back to the list of all Acanthamoeba species. Click on *Acanthamoeba pustulosa*. This one is clearly not so popular with the genome centers. Open the one Entrez Nucleotide record. The sequence is for the small subunit ribosomal RNA. If you were to go down the taxonomy list, opening all of the records, you would find large numbers of species with just this one gene sequence.

Why sequence the small rRNA? Copy the Accession Number for the *A. pustulosa* rRNA gene (AF019050). Now, click on the **NCBI** icon and then **BLAST.** Choose **Nucleotide BLASTN** near the top left. Paste **AF019050** in the window and click **BLAST.** Click **FORMAT** when it appears. The results page, which appears as a pop-up, finds large numbers of *Acanthamoeba sp.* Clearly from the graphical display they only differ one from another by small amounts. Close this result and go back to the BLAST page. Run a new search with the same gene but this time choose Primates only in the Options Organism drop down menu. A quick look at the graphical display of the result shows many conserved domains, and a look at the hit list displays highly significant hits on all of the primates, including ourselves. Repeat the search but this time limit it to the Archaea. Again the display shows highly significant hits on conserved domains. If you scroll down the results window you will see that many of the hits are listed as 'uncultured.' We have the sequence for these species but we cannot culture them.

Open some of the links to accession numbers for the uncultured samples. You will see in the Title field of the ENTREZ flatfile that they have been identified from environmental samples as diverse as tropical estuaries and Yellowstone hot springs.

We can conclude from this exercise that small subunit ribosomal RNAs have highly conserved domains extending across all living creatures (or at least any we have ever found). We can use this information in the laboratory to design Polymerase Chain Reaction (PCR) primers that will amplify these regions from any organism's genome (so-called Universal Primer sets). Researchers interested in microbial diversity isolate DNA from environmental samples such as soil or water then use PCR with these primer sets to amplify and clone segments of the ribosomal genes from organisms that were present. Go back to Taxonomy, Tree, Eukaryota. Scroll down a page or two and you will see a long list of Uncultured marine alveolates. Open a few of the records and notice that the one ENTREZ entry for each is the small subunit rRNA.

Clearly, if we have sequence data from an organism we can use it to identify species and build phylogenetic trees based on alignments. In environmental genomic samples we often find sequences that do not match any previously identified organism or group. We are defining new species based on these data. We would like to find the sequence more than once to reduce artifacts of sequencing mistakes. With sufficient data and consistent results we can define a putative species as those showing one or two percent sequence divergence. If it comes from a very rare species, then we may never see the organism.

At the top of the page toggle the Nucleotide field on, then click **GO.** Now you can see the number of ENTREZ nucleotide records for each species. For a large proportion it is only one and this is almost always rRNA due to its high degree of conservation.

Let's look at an example of an environmental genomics research paper. Open the link to a publication in the journal *Nature* by Giovannoni et al., 1990. (link: http://www.nature.com/nature/journal/v345/n6270/pdf/345060a0.pdf), which will appear in a pop-up window. This paper is an early example of such an rRNA-based environmental search. The environmental sample was from the Sargasso Sea in the center of the Atlantic Ocean, and Fig. 1 gives a phylogenetic tree of the sequences found. The surprising result was that almost all of the species found had never previously been described. They are simply labeled as SAR plus a number for the isolate. Note also in Fig. 2 of the paper a diagram of the secondary structure of a small rRNA gene displaying the regions sequenced and the positions of the sequence variation found in the samples. If you have time you might enjoy looking at a second paper dealing with the hot springs at Yellowstone (Barns et al., 1994 http://www.pubmedcentral.nih.gov/picrender.fcgi?artid=43212&blobtype=pdf).

The conclusion from many environmental samples is that most (perhaps 99%) of all microbial phylogenetic diversity is related to organisms that have never been cultured or described. Extensive sequencing is resource intensive. Sometimes researchers will categorize samples by the presence in the sequence of a restriction endonuclease cut site characteristic of a particular group of related organisms. This is basically restriction fragment length polymorphism analysis on the cloned isolates and you would have heard of that in your genetics classes. Such a strategy provides a snapshot or profile of the types of organisms in a sample although usually not with species level identification or information regarding the full breadth of the diversity. It is fast and cheaper though, and that also has a place in science!

Clearly in taking a random sample of organisms from the soil and amplifying their rDNA, the frequency with which the various species are present in the sample influences whether they will be detected. Rare species are only found if large numbers of sequences are analyzed.

Recent very large scale analyses have disclosed a hitherto largely unexpected microbial diversity and enormously enriched the NCBI databases with new sequences. Large genome centers have enormous analytical capacity and some recent environmental genomics studies have generated literally millions of new records. Let's look at one of these.

The strategies so far have focused on the conservation of a particular highly conserved gene but a result depends on our ability to specifically amplify and sequence it. An alternative strategy would be to isolate environmental DNA samples and simply to randomly subclone all of it. The clone inserts are then sequenced and then analyzed. Typically this continues so long as the researcher can afford it or until the analysis appears to be saturated. No sample has been saturated yet! This general strategy is the same one used in the so-called shotgun assembly strategy of whole genome sequencing.

In 2004, samples of Sargasso Sea water were analyzed using this strategy. A staggering one billion base pairs of unique sequence was generated (Venter et al., 1994 http://www.sciencemag.org/cgi/reprint/304/5667/66.pdf). In this type of study the sequence is from all regions of the genomes present in the samples, not just the rDNA. Nevertheless there was a remarkable observation that a high proportion of the sequence generated was only found in single sequence runs. The mathematics of such random sampling strategies tells us that when a large proportion of the observations are not replicated then the sampling is very far from saturation. More sequencing would have found many more species. The dataset suggested at least 1800 species were present and identified 1.2 million new gene sequences. This number of new genes was larger than the sum of all previously annotated genes in the databases. Subsequent studies suggest that the rare biosphere diversity in the oceans of the world may be much larger yet.

Apart from exploring biodiversity for its own sake, how else might we use this approach? Environmental samples over time show that microbial diversity can be altered by environmental changes such as climate change or environmental pollution. Once we have identified the species in a system, we can design DNA microarrays that efficiently identify the species in the samples. This provides a method for environmental monitoring. Changes in the species composition can be correlated with conditions such as pollution.

The field is not limited to sea water or soil. The strategy is being applied to human and animal health. For instance, the approach has discovered previously unexpected microbes in the oral cavity and gut. Changes in the distribution of the microflora and fauna may provide sensitive measures of health and disease.